Fe. 19 1994

To Howard Ris,
With the author's compliments

Thomas Lee

6B
HLTH
OTHER
93

7721 679

GENE FUTURE

The Promise and Perils
of the New Biology

GENE FUTURE

The Promise and Perils
of the New Biology

Thomas F. Lee

PLENUM PRESS • NEW YORK AND LONDON

Library of Congress Cataloging-in-Publication Data

Lee, Thomas F.
    Gene future : the promise and perils of the new biology / Thomas
  F. Lee.
       p.   cm.
    Includes bibliographical references and index.
    ISBN 0-306-44509-3
    1. Molecular biology. 2. Genetic engineering. 3. Molecular
  genetics. 4. Medical genetics. 5. Forensic genetics.   I. Title.
  QH506.L35 1993
  575.1--dc20   93-27834
                                                                 CIP

Illustrations appearing in this volume were prepared by Sally Green.

ISBN 0-306-44509-3

Printed in the United States of America

ACKNOWLEDGMENTS

I thank my many colleagues who offered their advice and encouragement during the writing of this book. I am particularly grateful to Karen Metz, Sally Field, Susan Martin, Jeanne Welch, and Eileen Lowry-Whittle of the Geisel Library staff for their gracious assistance. Thanks as well to Jim Vitale from TSI in Worcester, Massachusetts, for his advice on the subject of transgenic animals, to my daughter Julie for her untiring help with reference materials, to Sally Green, a talented and meticulous illustrator who prepared the artwork for this volume, and to Peter Heywood from Brown University, who supplied me with very helpful information on biotechnology and Third World agriculture. My work could not have been completed without the loving generosity of my wife, Eileen, and my children, Anne, Brian, Emily, Julie, Bridget, and Rebecca.

CONTENTS

GENE FUTURE
The Promise and Perils
of the New Biology

Chapter 1

ENTER THE GENES

On April 25, 1953, nine hundred words changed the world. Those words constituted the brief report by James Watson and Francis Crick that appeared in the renowned science journal *Nature*. They began, "We wish to suggest a structure for . . . DNA. This structure has novel features which are of considerable biological interest." They ended as modestly: "It has not escaped our notice that the specific pairing we have postulated immediately suggests a possible copying mechanism for genetic material."

Years later, in 1991, Watson would write in the second edition of *Recombinant DNA*, "There is no substance so important as DNA . . . the key to our optimism that all secrets of life are within the grasp of future generations of perceptive biologists is the ever accelerating speed at which we have been able to probe the secrets of DNA."

The brief *Nature* article, written when Watson was a 25-year-old postdoctoral student at Cambridge, is regarded as a masterpiece of understatement. That can hardly describe the words Watson, by then a

Nobel laureate and world-renowned scientist, wrote almost four decades later.

How could a description of the structure and workings of DNA—a mere molecule—evolve into a claim that DNA is a treasure map leading to a grasp of "all the secrets of life"? The full details of the map were promised only to "perceptive biologists" of the future. But already, Watson and Crick's analysis of that startlingly simple and beautifully symmetrical molecule has revealed precious secrets residing in the very control center of life—the *genes*.

Earth's biological past is a four-billion-year journey of the evolution of living organisms, all of which, in each of their cells, harbor long, densely coiled strands of DNA molecules. These serpentine coils are gathered into discrete bundles, the *chromosomes*. Along the length of each chromosome, the genes (perhaps as many as 100,000 in a human being) are each a unique segment of the chromosomal DNA. The precise details of the molecular structure of the DNA in each gene spell out a *genetic code*. The cells "read" the code and respond to its commands. And the result? Life—in all its marvelous forms and functions.

The intriguing story of the search for what lies at the core of the unique abilities and activities of living things goes back to the earliest days of science. It has always been obvious that living creatures have special qualities—growth, movement, reproduction—that set them apart from everything else. All forms of life appear to generate others of the same kind. How can each creature be unique and yet share in this mysterious common denominator, life?

As the search has widened, and the tools of science have become more sophisticated, scientists have been able to delve ever deeper into the inner recesses of cells. They finally arrived at the cell *nucleus*, the membrane-bound sac enclosing the chromosomes. Going further, they learned how to extract and study the chromosomes themselves. The chromosomes turned out to be fashioned out of DNA and proteins. How could either of these molecules hold the key to both the unity and the diversity of life?

When Watson and Crick presented their evidence for the structure of DNA to the scientific community, they did much more than describe the three-dimensional symmetry of just one of the thousands of chemicals found in living cells. They had uncovered the secret which would soon lead to a "New Biology," in which the evolution and abilities of each living creature would be seen precisely as the result of information flowing from its genes—messages carried not in the nuclear proteins but in the molecules of DNA.

The future of the biological sciences quickly centered around genes. The studies would be far from academic. The applications of this new approach are already beginning to reach into almost every facet of our lives in ways which range from life-saving gene-based diagnostics and treatment to the fashioning of hitherto impossible new genetic forms of animals and plants. Scientists are already well under way on the Human Genome Project, a 15-year-long effort to construct a map locating the 100,000 or so genes which are spread out over our 23 pairs of chromosomes and which spell out the genetic directions for the human race.

We have already entered the first stages of what can be termed a "Gene Future." It promises us a new and unprecedented understanding of genes, offering a novel and illuminating view into the most intimate operations of living things from microbes to humans—and with that, the potential to gain increasing control over life itself. It is projected to have an enormous impact on medicine, agriculture, industry, law, and the environment.

But there may be perils along that path as well. Are there dangers in this technological approach to nature? What may be the chances of unwittingly doing irreparable harm to individuals, societies, or the rest of the biosphere? What controversies have already arisen over possible perils as molecular science moves out of laboratories and into our clinics, fields, farms, and families?

Knowledge may not always lead to power, but it is certainly a critical prerequisite to making an informed decision about issues which surface where science and society meet. We are facing a change which could surpass the Industrial Revolution in its impact on the world. We hope that the succeeding pages will contribute to a deeper understanding of the fascinating story of the New Biology and assist in the decisions we are already being called on to make.

The terms *gene* and *DNA* have been imprinted on our national consciousness. Hardly a day passes when the words do not appear in a

headline or article in a newspaper or other periodical. We read that the
genes for diseases like cystic fibrosis or a particular form of cancer have
been "found." There are reports of legal tussles over whether or not
courts should accept the evidence of "DNA fingerprinting." Popular
accounts of articles published in current medical journals speak of
experimental "gene therapy" protocols being approved for use in humans. "Genetically engineered" plants will soon be available in our
markets, while down on the "pharm" such "genetic engineering" is
prodding animals to produce pharmaceuticals for medical treatments.

There are likewise stories recounting the voices of concerned
individuals, organizations, or even governments questioning, criticizing,
or sometimes even condemning these activities. Some of these critics
warn that if the genes are, in fact, at the core of our living systems, do we
dare tamper with them? Do we put ourselves or others at unacceptable
risk? Are we, to use a well-worn phrase, "playing God"?

The scientists who gathered in February 1975 at Asilomar, the
picturesque California conference center located at the edge of the
Pacific, were perhaps concerned not as much that they were "playing
God" as about the widespread impression that they were playing with
fire. The 139 researchers, along with a number of science administrators,
journalists, and experts in law and ethics met in response to an appeal
from the National Academy of Sciences, a prestigious independent
organization, which has advised the government of the United States
since its charter was approved by President Lincoln in 1863.

In April of 1975 a committee of respected Academy scientists had
called for a worldwide moratorium on certain types of gene research. The
restriction included aspects of research with tumor-inducing viruses and
with genes responsible for producing potent toxins in bacteria. They were
concerned over possible hazards to the scientists, their co-workers, and
the public at large. The moratorium was universally honored. Gene
scientists, more accurately referred to as *molecular biologists*, while
at odds over the degree of danger in the research under question, agreed
nearly unanimously that some controls were appropriate at that early
stage.

Their recommendations led to the formulation by the National
Institutes of Health of a set of strict gene research guidelines that took

effect in July 1976. As a wide range of molecular DNA research continued with no ill effects, scientific and public concern abated. By 1979 far fewer restrictive regulations were put in place. As James Watson so laconically put it, "the[y] [the original guidelines] took force in July 1976. They remained in effect until early in 1979. When they were relaxed, effective work on cancer viruses commenced."

Within less than three decades after Watson, Crick, and colleagues had first deciphered DNA, that once enigmatic molecule, core of our chromosomes and the stuff of our genes, had become one of the most powerful tools of science. As techniques emerged to read even the fine print of the molecular messages written into the DNA of all living things, we began to trace in the history of those messages new and exciting evidence of our origins and evolution. Genes were isolated from plants, animals, microbes, and humans and copied, multiplied millions of times over in the laboratory.

The extracted and amplified genes were inserted into the cells of living organisms sometimes far removed from any natural relationship with the source of those genes. Bacterial genes were insinuated into yeast cells, or ended up in amphibians, and animal genes found themselves being put into plant cells—and in many cases the newly arrived genes functioned perfectly, as though they were still in their natural home.

These molecular machinations were dubbed "genetic engineering." The possibilities for productive permutations of life seemed endless to the "molecular engineers," biologists who succumbed readily to the fascination of entering and rearranging the internal workings of living cells.

How could a single molecule, DNA, prove so alluring to scientists and offer them such power? The answer lies in the molecule itself and its central role in that unique complex of phenomena we call life.

*   *   *

C. P. Snow, the late British scientist and writer, in his 1959 classic *The Two Cultures* had vigorously championed increased communication between scientists and nonscientists. He wrote with a single-minded intensity that not only should we increase the scope of this communication but that, in fact, "the worst crime is innocence." Surprisingly, in a sequel written in 1964, long before any actual genetic engineering had taken place, Snow added: "I should now put forward a branch of science which ought to be a requisite in common culture. . . . this branch of science at present goes by the name of 'molecular biology.' "

While we are following Snow's advice, this introductory chapter is by no means intended to be detailed enough to be more than the slimmest of primers on molecular biology. However, we do need to head together into the succeeding chapters armed with the common denominator of a vocabulary describing genes, the DNA out of which they are fashioned, and their pivotal role in living systems.

But first, as we begin this book's journey through the New Biology's recent accomplishments, its highly publicized and promised potential, and the public debates swirling about its more highly controversial applications, it is also important to point out one aspect that we have chosen not to emphasize in this book.

Biologists study life most effectively when they approach their subject with the assumption that, given enough time, equipment, and perseverance, one can eventually learn all there is to know about it by understanding its building blocks, from muscles to molecules, and how these act in an integrated way.

This has been a most effective working assumption. Its rigorous application has permitted modern science to conquer smallpox; to defeat polio, tetanus, and tuberculosis; to control the electrical system of the human heart; and to probe the messages in the chromosomes. But biological systems are made of chemical molecules: proteins, carbohydrates, fats, and so on. And so the study of biology eventually arrives at chemistry.

But molecules consist of atoms, and atoms consist of further subunits: protons, neutrons, electrons. Beginning with a living organism, whose anatomy and behaviors can be rather exhaustively described (and perhaps regulated), we delve deeper and deeper through the realms of chemistry until we are left with clouds of electrons and find ourselves in the discipline of physics.

Can we study the properties of living organisms by finding out the structures, functions, and interactions of their constituent parts, down to their very molecules? On a purely pragmatic level, biologists have no other choice. But as famed biologist François Jacob put it succinctly:

"Biology can neither be reduced to physics, nor do without it." When biologists speak of seeking the code within our genes as a key to how humans function, they are not necessarily doing so with the intention of denying the reality of what are still deep mysteries, such as human creativity, the longing for justice, or the stirrings of love.

They are simply working within a tradition which has answered more crucial and highly significant questions about living systems in the last hundred years than any other system of scientific analysis in all of human history. The most recent expression of this tradition, the study of genes, is now the mainstream of modern biology.

In examining genes as life's controlling molecules, must we accept the thesis that, for example, all human behavior, and by extension all human society is controlled inflexibly by a continuum of molecular determinants from the genes to the person to all of society? Such a proposition is at the center of current and heated debates, ably represented by such books as Richard Dawkins's *The Selfish Gene*, E. O. Wilson's *Sociobiology*, and *Not in Our Genes* by R. C. Lewontin, Steven Rose, and Leon J. Kamin.

In this book, we defer from entering that arena. Our specific intent is to describe our recent entry into what will surely be a "Gene Future," and how this impinges directly on our lives and health, and on the environment which supports all of life. While we will in various instances point out where the implications of a Gene Future are not only physical but philosophical as well, we leave it to others to argue the metaphysical merits (or demerits) of science's fixation on genes.

Our story begins with DNA.

THAT MIGHTY MOLECULE

Within each living cell there is at least one chromosome. The simplest and most ancient forms of life, the bacteria, lack a nucleus but have a single, central chromosome and often, scattered elsewhere in the cell, some *plasmids*, tiny rings of DNA bearing only a few genes. As we shall see, the tiny plasmids play a large role in molecular biology.

In human cells, each nucleus contains 46 tightly packed, densely coiled chromosomes (see illustration). If the chromosomes from but one cell were unraveled and laid end to end they would form a fragile string more than six feet long—but only 50 trillionths of an inch wide. It is a string of DNA, the same molecule which makes up the chromosomes of all living things. The specific arrangement of the parts of that DNA determine the characteristics of the cells which house it, as well as the characteristics and capabilities of the organisms built from aggregations of those cells.

The chemical identity of the parts of DNA were known years before Watson and Crick published their epochal paper describing how those parts fit together. Theirs was a work of synthesis. They pieced together the clues obtained by the experimental laboratory work of others—such as the Columbia University biochemist Irwin Chargaff, or the X-ray analysis of DNA by Rosalind Franklin carried out in Maurice Wilkins's laboratory at King's College, Cambridge.

| DNA | | Coiled | | |
| Double helix | Nucleosomes | nucleosomes | Chromatin | Chromosome |

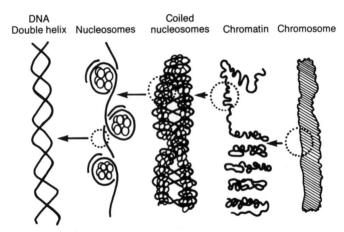

The *chromosome* actually is a very tightly coiled DNA molecule. The DNA is wrapped around clusters of proteins. This *nucleosome complex* is further coiled into *chromatin*, which is large enough to see with a microscope. When the cell is ready to divide, the chromatin is packed into short rod-like chromosomes.

The DNA model that Watson and Crick presented to the world in 1953 uncovered a symmetry which became the key to understanding how this molecule could account for the extraordinary diversity of living organisms as well as act as the control center for the complex inner workings of each of their cells.

Life (philosophy aside) is characterized by thousands of rapid, integrated chemical reactions called *metabolism*. What distinguishes the chemistry of life from that of nonliving systems is, in a word, organization. A cell, while in a sense nothing more than a sea of chemicals, is nevertheless able to grow, maintain, repair, and even reproduce itself. The conductors of this chemical symphony are the enzymes, proteins which act as catalysts to induce molecules to form new combinations or to allow more complex molecules to break down into smaller components.

All living things share many of the same chemical reactions and enzymes. The eons of evolution on Earth have been very conservative in the sense that many of the very same cellular chemical reactions are carried out by all living things. That is why such humble organisms as flies, yeast, or worms continue to yield valuable information directly applicable to questions about human metabolism.

That which makes one species of organism different from another related species, or one human different from another, can be stated simply: the cells of each make a different complex of enzymes. All enzymes are proteins. What specific proteins a cell makes is determined by a code in the DNA of its chromosomes. If you and I have the same DNA code, we are identical twins. The enzymes drive their specific, unique chemical reactions. The result? Life, in all its myriad forms and functions.

I may wish to have red hair, but unless my DNA code spells out the instructions for making the enzymes necessary to drive the chemical reactions in my hair follicles to synthesize red pigments, I wish in vain. Of course, given the proper code, my follicle cells would have little choice but to obey its directions, and I would have to endure being called a "carrot-top." The ability to transfer DNA codes from one organism to another is the basis of genetic engineering and much of biotechnology, to which we shall return in just a moment.

DNA is a deceptively simple, gently twisted chain of many thousands of repeating units of only four kinds—the *nucleotides*. Each of these consists of a phosphate group, the sugar deoxyribose, and one of only four possible nitrogenous (nitrogen-containing) bases—adenine (A), guanine (G), cytosine (C), or thymine (T).

The entire DNA molecule (see illustration) is shaped like an evenly twisted ladder, forming the famous double helix. (A helix is the shape which a string would form if it were to be wrapped around a cylinder, as opposed to a spiral, which would result if the string were wrapped around a cone.) In our analogy we will substitute a flexible ladder for the string. The sidepieces of the ladder are made up of alternating sugar and

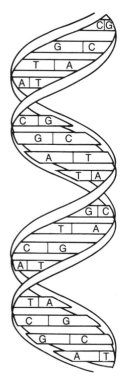

The famous DNA double helix "ladder" consists of two "sidepieces"—chains of alternating deoxyribose sugar and phosphate groups (here shown as two solid, twisting strands). These coiled chains are connected to each other by the "rungs"—pairs of nitrogenous bases. The pairs are always adenine (A) and thymine (T) or guanine (G) and cytosine (C).

phosphate groups, with the rungs consisting of nitrogenous bases attached to the sugars.

The rungs are actually pairs of bases loosely attached to each other across the narrow gap between the sugars in the sidepieces. Due to the regular dimensions of the twisted DNA "ladder" only A can bind to T, and only G can bind to C. Therein lies the answer to the mystery and power of DNA. The *genetic code* is imprinted in the precise sequence of the nitrogenous bases running down the length of the DNA molecule. This code, finally deciphered in its entirety by the late 1960s, consists of a series of base triplets, ATT, GCC, GAT, or other triplet combinations of the four nitrogenous bases.

But how can such a brief code contain enough information to operate one cell, let alone the trillions of cells in a human being? The answer lies in the fact that these base triplets are a code for making a vital substance, the cell's proteins. Protein molecules are long chains of *amino acids*. There are twenty chemically different amino acids in human proteins. When attached end to end in a specific sequence of sometimes thousands of amino acids, and then folded up like complex pretzels, these amino acid chains form the hundreds of different enzymes and other proteins needed by our cells. If the code is present, so will be its resulting proteins.

The three-dimensional shape of a protein is critical to its function, and sometimes a change in the position of a few or even one amino acid can change the nature of that protein. For example, a difference in one amino acid in the hemoglobin molecule in red blood cells causes a deficiency in its oxygen-carrying capacity (see illustration). The result is the devastating genetic disease sickle-cell anemia. The code in the DNA is scrambled and the protein produced is therefore abnormal.

The DNA code is translated when the cell undergoes *protein synthesis* (see illustration). The DNA double helix momentarily separates, and certain areas of the base triplet code (ATT, GCC, etc.) are exposed. These are used as a guide for the rapid synthesis of an intermediary to carry that code from the chromosome to the *ribosomes*, small, roughly spherical bodies on which the amino acids are stitched together to make the proteins.

The intermediary is *messenger RNA* (mRNA). This is a single-

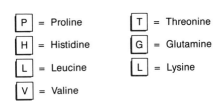

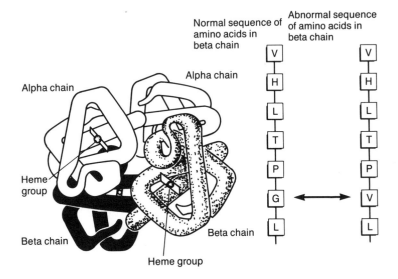

Hemoglobin molecule

The hemoglobin molecule consists of four intricately twisted chains of amino acids. Note that the difference between normal hemoglobin and that found in the blood of people with sickle cell anemia is merely the substitution in the latter of the amino acid valine for glutamine. This is sufficient to decrease the capacity of the red blood cells to carry oxygen to the body's tissues. (From David Suzuki and Peter Knudtson, *Genethics: The Clash between the New Genetics and Human Values*, Harvard University Press, 1989. Reprinted by permission of Harvard University Press. ©1989 New Data Enterprise and Peter Knudtson.)

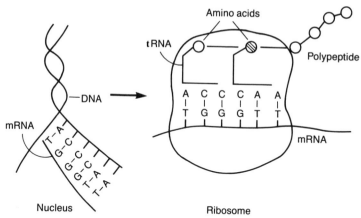

In *protein synthesis*, the messenger RNA (mRNA) is formed as a strand complementary to one side of the DNA helix. It moves to the ribosomes, to which *transfer RNA* (tRNA) brings amino acids. The latter are attached to each other in a sequence corresponding to the sequence of nitrogenous bases in the mRNA. The resulting strand, a *polypeptide*, will be joined to other polypeptides to make a protein. In this way, the code in the DNA is translated into a protein with a specific order of amino acids.

stranded molecule (continuing our analogy, one side of a ladder with half-rungs) in contrast to DNA, but still made from nucleotides. Messenger RNA forms along one or the other side of the opened DNA. The code of the base sequence of the DNA becomes incorporated into complementary (that is, matching) bases in the mRNA.

The strands of mRNA move away from the chromosomes, the helix closes, and the cell's ribosomes take over, literally moving along the length of the mRNA. The message transcribed earlier from the DNA is translated. The sequence of mRNA base triplets determines the exact sequence of amino acids being stitched together one after the other to make a protein. The message originating in the DNA and carried by mRNA to the ribosomes has been delivered, received, and understood. In summary, DNA makes RNA, which makes protein—a crucial process labeled by Francis Crick as biology's "Central Dogma," and a central event in the chemistry of life.

GENES AND GENOMES

What actually are genes? How do they fit into this story of life's chemistry? For our present purposes, a gene can be considered as any segment of a DNA molecule that contains the code for a particular protein. Considering the large size of most proteins, a gene may contain thousands of bases. There are somewhere between 50,000 and 100,000 genes in each human cell.

Chromosomes (except in bacteria) come in almost identical pairs. The full complement of chromosomes in a cell is referred to as the *diploid* number, and half that, as would be found in a sperm or egg cell, is the *haploid* number. An organism's *genome* is the totality of all the DNA in each cell of that organism. There are about three billion base pairs in the haploid genome of a human being. (In comparison, the haploid genome of a mouse also has three billion, while that of corn has 5 billion. It's the sequence that counts!)

This means that if the average gene is about 1000 bases long, and we humans need about 100,000 genes, about 100 million of our three to four billion base pairs make up our genes. In fact, almost all of our DNA does *not* contain codes for making proteins at all. There is much controversy over why all this "extra" DNA (some have called it "junk" DNA) has been retained in evolution.

Whatever the answer, we do know that at any one time in a given cell, only relatively few genes are actually functioning, or, as molecular biologists put it, are "turned on." After all, the genome of one diploid cell has all the information to make a complete human—and in the case of a fertilized human egg, it does just that. Consider, for example, cells in the retina of the eye. They must make pigments, the colored chemicals which react to the incoming light and make visual reception possible. In contrast, cells in the muscles of the arm must make proteins enabling them to contract.

In both types of cells there are exactly the same kind and number of chromosomes. Obviously, not all the genes in these differing cell types are turned on. If they were, chaos would result. In each cell, therefore, depending on where it is located in the body, some genes function while most others do not. Relatively little is known about how this process is regulated, but an intense search is under way. For when we understand how genes switch on and off, we will understand cancer.

We have known the three-base code for the position of each amino acid in proteins since 1966. It stands to reason that analysis of the amino acid sequences of proteins extracted from cells could tell us which genes are present in those cells. But the ultimate reading of the DNA code would require the spelling out of the entire nitrogenous base sequence of the DNA molecules making up the cell's chromosomes.

To succeed in such an ambitious undertaking would result in what would amount to a printout of base sequences made up of only four characters, A, T, G, and C. A complete printout for the human genome would be enormous—three billion letters—and would include not only the genes but the other 98 percent or so of "extra" DNA.

This printout is precisely one of the goals of the Human Genome Project. This international effort has so far determined the sequence of about 0.2 percent of the human genome. When the sequence is completed, simply printing the letters will require the number of pages in at least 13 sets of the *Encyclopaedia Britannica*.

More immediately relevant than the entire sequence is the progress being made by the participants in constructing chromosome maps. The object of mapping is to determine as accurately as possible the location of specific genes on particular chromosomes. By now, more than 2500 human genes have been tracked down to their chromosomal positions.

And why is that important? Among other reasons, there is a quite pragmatic and vital use for such information. Thousands of diseases—among them cancer, diabetes, cystic fibrosis—have a genetic basis. Mapping enables us not only to diagnose the presence or sometimes even the tendency toward many human genetic diseases, but also allows us to track down, isolate, and study those genes, in the hope of finding a cure or preventative.

The background and the ongoing drama of the Human Genome Project is a story in itself. In the following chapters we will concentrate on some of the spin-offs of that ambitious undertaking, as well as those from the mapping and sequencing of other organisms, both plant and animal. Our Gene Future already has many genes on hand, with very many more soon to be captured.

BIOTECHNOLOGY

Soon after the complete genetic code had been deciphered, extraordinary developments occurred which would move molecular biology from the laboratory to the marketplace. In 1970 scientists discovered that certain bacteria were rich sources of extraordinarily useful enzymes. These proteins, termed *restriction enzymes*, can act like molecular scissors and cut DNA molecules into discrete pieces. Then, in 1972, the bacterial enzyme ligase was shown to act as a kind of biological glue, which could rejoin those pieces. The availability of these enzymes soon allowed molecular dicing and splicing in which DNA fragments cut out of the genomes of organisms were stitched into the genomes of others.

DNA from organisms as far removed from each other over the eons of evolution as bacteria and humans could be combined into *recombinant DNA* (rDNA). Such "genetic engineering" had invaded and taken over life's control center. The stage was set for hitherto unimagined manipulation of living organisms.

Remember, the genetic code is (with a few exceptions) universal. That implies that a gene for luminescence from a firefly, if added to the genome of a carrot, should function and produce light. That, in fact, has been done, and carrots which dutifully glow (albeit rather weakly) have been constructed. That is of some technical interest, but hardly helpful to farmers, who seldom harvest in the dark.

However, imagine, for example, cutting out the human genes responsible for making insulin and splicing them into the DNA of a bacterium. The bacteria, which can be grown in enormous numbers, could then pump out insulin, obeying the orders of their newfound genes.

This is now done on a commercial basis, and people with diabetes routinely use this recombinant insulin.

Biotechnology took on a new meaning and focus with this novel power to use DNA as a tool to make marketable materials. Modern biotechnology can be defined in its most general sense as the scientific manipulation of organisms, particularly at the molecular genetic level, to produce useful products. In a sense, biotechnology is one of the oldest industries in the world. The fermentation of wine or the preparation of bread, both brought about through the metabolic activities of yeast, come under its rubric, as do the centuries of cross-breeding and hybridization of crops and domesticated animals.

Now, however, the limitations of the random mixing of genomes of related organisms through nature's time-honored process of sexual reproduction could be overcome. All that was needed was an ample supply of specific genes with known capabilities and a way to get them into the genome of, let us say, a tomato plant, a cow—or a human being.

Chromosome mapping and other techniques can pinpoint genes. These can be removed from any cells and put into bacteria, where they are copied by the millions as the bacterial cells divide. The amplified genes can then be extracted, purified, and, using rDNA methods, placed into the genomes in the cells of an organism of choice. Even this slavish copying of genes (called *gene cloning*) by bacteria can now be surpassed by an even faster process: The polymerase chain reaction, or PCR, can start with a few tiny fragments of DNA and copy it billions of times in a few hours—in an apparatus no larger than a microwave oven.

Biotechnology companies have sprung up by the hundreds over the last decade with all the attendant hope and hype one would expect from an industry with such enormous potential to make and sell the unique products made by genetically altered organisms. A few of these are products already available—such as insulin or human growth hormone made by genetically engineered bacteria. Our Gene Future soon will offer us a cornucopia of products made by other organisms as well—plants, animals, or even human cells—for scientific, medical, agricultural, and other uses, products which will influence the lives and health of all of us.

Molecular biologists learned how to alter the genomes of bacteria as early as the 1970s. Only in the last few years have molecular techniques reached a level of sophistication which has allowed scientists to turn their attention to organisms larger and more familiar to us, and these are the organisms on which we will focus. Designer genes are now being added to the chromosomes of creatures from rice to mice, as well as to those of humans in early trials of gene therapy.

We are witnesses to and participants in a watershed of human history. The evolution of life on this planet, from its humble beginnings about four billion years ago, has only recently resulted in creatures unique among all other species. We humans, self-categorized as *Homo sapiens*, "wise man," can trace our intelligence as well as all our other capabilities, in the final analysis, to the messages in our genes.

And now those genetic capacities, expressed in the ceaseless curiosity of scientific investigation, have permitted the paradox of organisms who have not only discovered the whereabouts of genes, but can literally use them to become the partners of mere chance in evolution.

Let's look at how we are beginning to scrutinize genes in order to read the very history of that evolution, to alter plants and animals in ways undreamt of only a few years ago, and to use genes as tools to detect and deter disease. In all of this there are promises to ponder, both realistic and inflated, and dangers to consider, both real and imagined. We have already entered the doorway of the Gene Future.

Chapter 2

GENES, GENEALOGY, AND ANCIENT DNA

Planet Earth, our home, is but a fragment within one of the hundreds of millions of galaxies which swirl about in the universe. Earth is one of nine planets orbiting our sun, itself only one of billions of stars in the Milky Way. The star nearest to our sun, Proxima Centauri, floats 40 trillion miles away, its faint light finally reaching us four years after its departure.

Twenty to thirty billion years ago, following the "Big Bang"—the cataclysmic explosion of an unimaginably dense, tiny core containing all that the universe encompasses—forms took shape in the crucibles of atomic fusion and condensation. Earth emerged as a lifeless sphere some 4.6 billion years ago, drawn together like its fellow planets from peripheral material left spinning around the young sun.

As warmth slowly welled up into the earth's surface from pressure and radioactive decay, molten material settled down into layers of varying density, the least dense solidifying on the surface as a thin crust. The primitive atmosphere of hot hydrogen gas boiled away, while volcanoes and other vents belched forth gases: carbon monoxide, carbon dioxide, nitrogen, methane, and ammonia. As Earth, bathed in withering ultraviolet radiation, once again cooled, torrential rains formed the seas to the roar (had there been ears to hear) of savage lightning.

And life began. Within a billion years there emerged an entity made out of the molecules of Earth. It was walled off from its surroundings by a membrane through which it could permit vital supplies to enter and keep other substances out. Such a *cell* would be an integrated chemical "factory," one which could maintain its equilibrium and even grow. And most importantly, it could copy itself, divide, and become two.

A capacity for random change was built into the very fabric of this life. The inevitable changes often brought death, that is, a disorganization of molecular structure incompatible with living functions. But sometimes the changes were not lethal; in fact they enhanced the capacity of the life form to endure, and it multiplied. Living organisms spread out over the earth, changing by the sheer volume of their chemical reactions the chemistry of the water, soil, and air of Earth.

These new conditions, in turn, helped shape the ever-changing parade of life. But pathways that appear as if they might have been productive more often than not ended in oblivion. Epochal events such as geological upheavals, climatic changes, or drifting continents, as well as small, chance, isolated incidents—a mud slide, a diverted stream—led to extinction. Estimates of the progeny of evolution over the ages range from 5 to 50 billion species. Today there are an estimated 5 to 50 million survivors—and extinctions are occurring at an increasing pace.

We live in one of many possible worlds. From a long, meandering trail of evolutionary changes finally emerged *Homo sapiens*, who appears to be, as paleontologist Stephen Jay Gould so piquantly expresses it, "a tiny twig on an improbable branch of a contingent limb on a fortunate tree."

There was born in *Homo sapiens* a ceaseless curiosity which, imprinted in the primitive brain and applied over the centuries, ultimately gave rise to modern science. Perhaps for the first time in the long history of the universe aggregations of living cells, human beings, were able to use specialized neural tissues to ponder the mysteries of their own functions and origins. Our genetic endowment has enabled us literally to turn and look at those very genes and in that inquiry discover secrets of the innermost control of life and ourselves.

What have we found in that search for our history? What do we know of our ancient human ancestors or the progenitors of our species? How can we possibly trace our lineage back through the corridors of times so distant that the span of recorded history or even the duration of the existence of sentient beings is as a few seconds compared to a year?

TIME'S TRACES

Fossils are needles in the haystack of history. These preserved remnants of organisms lie scattered about the earth, some still where creatures perished or often far from the death site, carried by floods or upheavals which may have buried them to inaccessible depths or lodged them where they might eventually be discovered. In rare circumstances an entire organism might be preserved. More often, whatever organic material escaped decay might remain as a thin film compressed between layers of sandstone and shale.

Hard parts such as the bones or teeth of vertebrates or the shells of invertebrates may survive as such. Some organisms undergo petrifaction as minerals dissolved in groundwater slowly seep into the dead tissues and replace the organic material, turning cells and tissues into stone. Often the fossils found are not even petrified remains but replicas molded when sand or mud covered the scene of death.

The conditions permitting fossilization are chancy and the "fossil record," as it is called, is notoriously incomplete. The garnering of fossils is inherently limited by their sparse supply. Relatively few individuals compared to the number of devotees of other scientific disciplines have devoted their time to fossil collection. However, their expertise, drawn from experience in diverse scientific specialties, in interpreting what the fossils have to tell us has yielded a fascinating outline of at least some of the broad trends in the history of life. In some instances we have been privileged with some remarkably detailed glimpses of bizarre byways of evolution, now long extinct.

But despite the efforts of dedicated men and women to discover and read this record of the rocks, the fact remains that our desire to clarify even the problematic history of our own genus and species, let alone that of the countless organisms too small or delicate to be preserved, cannot be fulfilled by fossil study alone. Even when we supplement this knowledge by careful scrutiny of and comparisons among such measurable features as the gross and microscopic anatomy of extant organisms as well as their habits and habitats, we find that too much history has been lost. It seems that time has erased the records. But has it? We have turned to the genes and in them may have found the Rosetta stone of evolution.

THE BOOK OF LIFE

Few reactions are as predictable as those of my students when I introduce the subject of *taxonomy*, typically defined as "the science, laws, or principles of the classification of organisms." A pall descends on the classroom as they gird themselves for the expected recitation of a desiccated diary of names and categories. They fail (initially) to understand that the pursuit of taxonomy is the very essence of science itself.

Science is about curiosity, about finding explanations for the world around us. Biological science centers on the fascinating phenomenon of life. Scientists do not wander about, arbitrarily assigning names to organisms and placing them in categories for convenience. The essence of taxonomy is the effort to discern the real relationships among the millions of living beings that populate this planet. It is a search for nature's family archives, for who we are and how we and our fellow creatures came to be. It now appears that much of that history, long assumed to have disappeared forever, may still be written in the genetic code of every living cell.

In the cell's DNA, its RNA, and their protein products are passages, which we are only now beginning to read, filled with clues about life's very origins and subsequent evolution. By comparing these texts inscribed in surviving species we are able to peer into the comings and goings of creatures long since extinct, many of which were necessary precursors of ourselves. How is this possible, and what have we learned?

Systematics Goes Molecular

Within a decade of the startling news that the genetic code had been deciphered, Linus Pauling and Emile Zuckerkandl of the California Institute of Technology inspired a small but determined group of researchers to begin a paradigm shift in systematics. In 1962 and in more detail in 1965, Pauling and Zuckerkandl suggested that mutations—that is, changes in the nitrogenous base sequence of DNA and RNA—and consequent alterations in the amino acid sequence of the proteins whose architecture is determined by the DNA and RNA could serve as evolutionary "clocks."

They pointed out that the more mutations that had accumulated in a protein sequence, the older its lineage is. In other words, as these rare but inevitable variations occurred in DNA, perhaps during cell division or produced by outside influences such as radiation, the resultant messages copied into RNA and then into DNA would reflect those changes.

Moreover, the differences in these sequences could be used to compare living species to help determine their kinships. Closely related organisms could be expected to share DNA, RNA, and proteins of greater similarity than would be found among more distantly related individuals.

Of course, to use mutations as a clock one would have to assume that such changes occur at a constant rate. As we shall soon see, this assumption would sometimes prove problematic. But in the following case, it would prove to be productive and revolutionary.

By 1967 Vincent Sarich and Allan Wilson at the University of California at Berkeley had used what is now a relatively crude tool of molecular biology to analyze the differences in antibody responses among the blood proteins of primates (humans, apes, Old World monkeys, and New World monkeys). These alterations in their blood proteins can be traced back to changes in the genes coding for these proteins.

They now had a molecular clock. To calibrate it they calculated the mutation rate in other species whose history could be clearly inferred from fossil species. Their data led them to support the widely held notion that humans, chimps, and gorillas evolved from a common ancestor. This confirmed the conclusions published in 1962 by Morris Goodman, an immunologist at Wayne State University. Moreover, Sarich and Wilson's interpretation of their data led them to hypothesize that this common ancestor had existed about five million years ago, not the familiar 20 to 30 million years inferred by paleontologists from fossil evidence. (Wilson's pioneering participation in this research ended with his death from acute leukemia in July 1992 at the age of 56.)

The reaction from many established scientists was immediate and "vitriolic," in the word of Wes Brown, then a graduate student in evolutionary biology who in the late 1970s would introduce the techniques of DNA analysis to Wilson's lab, establishing it as the first to use DNA analysis to study evolutionary questions. Even Goodman was among the major dissenters.

Such a comparatively recent branching date would seem to leave far too few millennia for all of the alterations that would have to have taken place between early human-like creatures (the hominids) and our own genus *Homo*. After all, the major contender for the oldest hominid was the famous *Ramapithecus*, known from Indian and African fossils to be 8 to 14 million years old.

The Sarich and Wilson hypothesis had raised the possibility that the study of molecules might now be needed to enhance, or in some cases even replace, the familiar classic approaches of anatomical studies and fossil description as the tool for clarifying the history of life on earth. Would a painter discard brushes and oils for computer graphics or a surgeon a scalpel for a laser? The stage was set for an ongoing turf war between the molecular upstarts and traditional practitioners.

The proponents of the molecular school were at a distinct disadvantage. While their suggestions were intriguing, the techniques now familiar to us for exposing the intimate details of proteins and especially DNA and RNA were yet to be developed. Allan Wilson's attempt to revise so dramatically the story of human ancestry, coupled with the novelty of choosing biochemistry over bones, led many paleontologists to reject the molecular approach altogether—and a protracted struggle was under way.

It was not until almost twenty years later that Sarich and Wilson's radical position was finally vindicated. During the 1960s and 1970s Elwyn Simons and David Pilbeam, both then working at Yale, had concentrated their search for a hominid ancestor among the Miocene apes, whose fossils were scattered widely over Europe and Africa. A complete jaw had been reconstructed by Simons by forming a mirror image from a half-jaw of *Ramapithecus*. The teeth appeared to be arranged in a curve around the jaw, a human trait opposed to the rectangular rows of ape teeth. A rough estimate of the size of the animal could be made; it stood about three feet tall. No other details were known. *Ramapithecus* became a familiar member of the hominid line.

In 1976 another, larger piece of *Ramapithecus* jaw was found, which seemed to indicate that it was not as humanlike as had been thought. Further finds of other Miocene apes in 1980 and 1982 dealt a final blow to the honored position of *Ramapithecus* as one of our ancestors. Its descendants became not humans, but orangutans.

Early Life

Before continuing with our molecular search for the human family tree, let us first return to the very beginning of the story. The first forms of life on earth have long since disappeared. No witnesses were present, no fossils formed. But drawing on molecular clues retained in contemporary cells, chemists and molecular biologists have built models of what may have existed.

Throughout the 1960s and 1970s the Central Dogma of biology had become a familiar story—"DNA makes RNA makes protein." The genetic code in DNA flowed from DNA to proteins, with RNA the humble go-between molecule. Believers in this comfortable universal statement had to accept some exceptions with the discovery in the 1980s that RNA can indulge in the kind of sophisticated chemistry previously thought to be the exclusive domain of proteins.

In 1982 Thomas Cech at the University of Colorado at Boulder astonished his readers when he reported finding a segment of RNA in the single-celled protozoan *Tetrahymena* that could act as an enzyme. In

1983 Sidney Altman and his colleagues at Yale discovered another RNA enzyme inside the common bacterium *Escherichia coli*. In 1986 Robert Benne of the University of Amsterdam stumbled upon a remarkable phenomenon. While studying the genes of *Trypanosoma brucei*, a parasite that causes sleeping sickness in humans, he found that the parasite has a missing message in its genetic code, so that it makes defective RNA—but the defect is corrected by RNA itself. This and subsequent similar discoveries soon gave rise to the notion of the plausibility of a primitive "RNA world."

Perhaps at first RNA-based rather than DNA-based genes within cell-like systems coded for RNA enzymes, rather than the now-familiar protein enzymes. Later these evolved the biochemical machinery to translate DNA codes into protein enzymes, resulting in the common ancestors of all modern forms of life. The earlier, original, RNA-dominated forms vanished, leaving behind a few remnants of this earlier time. These traces are still with us, including some viruses, such as those that cause AIDS and influenza, which carry their deadly code as RNA rather than the DNA of most viruses.

How could such complex molecules as DNA and RNA have formed, survived, and prospered in the turbulent chemical soup of early Earth? RNA breaks down rapidly in water. Perhaps surface scums provided a refuge. Several other hypotheses have been suggested, including a provocative scenario detailed by Graham Cairns-Smith of the University of Glasgow. He posits that life might have begun with clay crystals that carried a primitive genetic code in the form of defects in the crystal structure. Such crystals might have provided a surface on which the synthesis of organic molecules might have taken place.

Chemist and historian Robert Shapiro has written:

> We must wait for the final answer concerning clay life. . . . It would be one of the most satisfying scientific answers. We inhabit this planet and use its resources. Our bodies are placed into the earth when we pass on. How fitting if we were ultimately born of this soil as well.

We may never know precisely how and when life emerged on our planet, but some four billion years later Earth teems with forms of life from bacteria far below the limits of resolution of all but the most powerful microscopes to great whales and expansive forests of towering trees. What is the family tree whose trunk and branches connect these diverse and unique creatures? And most especially, who are our relatives and ancient ancestors? How could rational beings conscious of self and insatiably curious about their history have emerged from all the rest? Again we have turned to the genes to help us find the answers.

All Creatures Great and Small

Every schoolchild is familiar with the categories of *plant* and *animal*. For centuries naturalists tirelessly cataloged living organisms as belonging to one or the other of these groups, called kingdoms, by analyzing their anatomy and lifestyles. With the advent of Darwinian concepts of evolution, taxonomists of the late nineteenth century were forced to refit the known organisms into classification schemes of "natural" units reflecting their evolutionary relationships, as far as they could then be determined.

As the fascinating world of microscopic life attracted increasing study, the myriad of tiny creatures invisible but for the lens proved too puzzling to label simply as plant or animal. In 1866 a third kingdom, the *Protista*, was proposed by the German biologist Ernst Haeckel to include these single-celled microbes. And so the three kingdoms remained until well into this century. By the 1960s scientists were using extraordinarily powerful microscopes in which beams of electrons probed subcellular details, revealing a major previously unknown distinction between one group of creatures and all other living organisms.

The bacteria proved to be unique. They showed surprisingly little internal complexity. They have only about 1/1000 as much DNA as all other cells, and the DNA is not surrounded by a membrane but lies naked in the midst of what appeared to be an almost featureless cellular landscape. Every other type of cell examined from worm to whale has a complex of *organelles*, variously shaped structures bound by membranes. Organelles include, for example, the *mitochondria*, sites of energy-releasing chemical reactions, and *dictyosomes*, where chemicals are stored and secreted. The bacteria, obviously very different, were labeled a superkingdom, the *Procaryotae* or *prokaryotes* ("before the nucleus"), and all other organisms were placed in the superkingdom *Eucaryotae* or *eukaryotes* ("true nucleus").

While science is open to change and growth as perhaps no other discipline is, usage breeds familiarity. Introductory biology textbooks in the 1970s and 80s presented a tidy taxonomic scheme. There were now five widely accepted kingdoms. These were the *Monera* (bacteria), the *Plantae* (plants), *Mycetae* (fungi), *Animalia* (animals), and the *Protista*, a pared-down but still enigmatic group consisting of microscopic, mostly single-celled eukaryotes, such as many of the beautiful, abundant, and varied plankton in fresh and salt water.

Students of the 1960s felt quite modern as they dutifully listed the textbook taxonomic schema. But by the late 70s, molecular biology had come into its own and with it the tools to peer even beyond the intricate scaffolding of cells into their molecules.

Reading the Ribosomes

Photographs of cells taken through the powerful electron microscope show their ribosomes as tiny spheres scattered about the cell or attached to membranes. A bacterial cell may house a few thousand ribosomes, while a eukaryotic cell may harbor millions. Their role is critical to life. The ribosomes serve as workbenches on which the synthesis of proteins takes place. As was noted in Chapter 1, here the genetic code arrives from the nucleus in the form of messenger RNA, which drapes itself over the surfaces of the ribosomes where its code is read and translated into the corresponding sequence of amino acids. These acids then fold into the intricate forms of proteins.

But on closer inspection, ribosomes are revealed as not perfectly spherical. They are made up of one larger and one smaller contorted subunit, each of which consists of a complex of proteins and a unique form of RNA, termed *ribosomal RNA* (rRNA). Not all genes are involved in protein synthesis; some code instead for rRNA. This means that the base sequence of a cell's rRNA is a reflection of the base sequences in some of that cell's genes. The proteins and rRNA in the ribosomes of prokaryotes differ greatly from those in the eukaryotes, pointing to the profound differences between their genes.

Most people tend to associate bacteria with disease. In fact, very few species of bacteria can do us harm. These invisible cells are virtually everywhere, silently breaking down organic matter in the water and soil, recycling nutrients as they release vital molecules into their surroundings. Global nutrient cycles (and therefore life itself) depend on the bacteria.

Some of these microbes flourish in unlikely environments. Hot sulfur springs, smoldering acid-laden tailings from coal mines, and the "lifeless" Dead Sea and Great Salt Lake have long been known to teem with unique bacterial species. When Carl Woese, a bacteriologist at the University of Illinois at Urbana, began to take apart the amino acid and RNA sequences of these kinds of bacteria in the late 70s, the neat scheme of life began to unravel. His and subsequent analyses revealed that these organisms, the *archaebacteria*, are as different from other prokaryotes as the prokaryotes are from eukaryotes.

By the 90s students were comfortable with the presence of the strange but mildly interesting archaebacteria tucked away with the rest of the bacteria among the prokaryotes, so beautifully illustrated in their new texts. But now, according to Woese's data, there was perhaps a third line of evolutionary descent. The archaebacteria (Woese prefers the term *archaea* to underline their uniqueness) should not be grouped with the bacteria at all. The prokaryotes had appeared first, at least 3.5 billion years ago. These soon gave rise to two divergent paths, one of which led away from the majority of the bacteria (now called *eubacteria*, or "true" bacteria) into the unique archaebacteria, which evolved specialized lifestyles. Much later (perhaps two billion years or so) primitive eukaryotes evolved from eubacterial ancestors.

Aided by increasingly available and more refined technology, scientists probed deeper into the ribosome's RNA and protein sequences. Publishers once more had to ponder the contents of their upcoming editions as evidence mounted that the now-familiar evolutionary trees should be uprooted. The genes were telling us that prokaryotes and eukaryotes diverged far deeper in the evolutionary history of life than we had ever thought possible.

They pointed to a history in which eukaryotes were as old or nearly as old as the eubacteria and archaebacteria. In addition, the extraordinary diversity of eukaryotic life, which numbers us among its members, is relatively recent, in the densely branching crown of the evolutionary tree. The notion of the five kingdoms became highly suspect. Evolutionary distances between plants, animals, and fungi and among the bacteria are dwarfed by genetic differences within members of the Kingdom *Protista*. Woese and his colleagues maintained that there are only three real superkingdoms or "domains," the archaebacteria, the eubacteria, and the eukaryotes, the latter including plants, animals, and other kingdoms.

Meanwhile, Woese's analyses have expanded far beyond that of bacterial rRNA. He and co-workers have undertaken to determine the base sequence of an entire bacterial genome, which encompasses more than two thousand genes. The project is expected to take at least ten years. Perhaps it will lead us to the roots of life—to the primitive genes from which all modern genes have descended.

Well, how many kingdoms are there? What is the true story? The answer is that in science, there is no dogma, there is only evidence. The data coming from the analysis of the genes may necessitate a dramatically revised view of life's history.

That remains to be seen. Whatever the outcome, one would hope that by the time that freshman biology textbook readers have become practicing scientists their initial discomfort at the interruption of familiar interpretations has long since disappeared. They should come to realize that growth in knowledge and true progress in the search for answers comes only when one is open to the possibility of the latest "truth" being discarded as new ways of understanding are uncovered.

OUR OWN STORY

Fossils may not be enough to tell an entire story, but the bones still have much to say. The best approach to our own history is now a collage of fossil and molecular evidence. As Maitland Edey and Donald Johanson so aptly say in their 1989 book *Blueprints: Solving the Mystery of*

Evolution: "There is a bone story to be told; then it must be edited by molecules."

The tantalizing clues about our history that we have derived from human and prehuman fossils have often been controversial. Charles Darwin would undoubtedly have been delighted to have had the opportunity to engage in such controversies. But while it had been clear to Darwin when he published his epoch-making treatise *The Origin of Species* in 1859 that humans must have evolved as surely as all other creatures, he could say nothing more edifying than, "Light will be thrown on the origin of man and his history." For no clearly human fossils had yet been found.

To be sure, a few limb bones and part of a skull of some kind of primitive human had been uncovered by workers digging in a quarry in the Neander Valley near Düsseldorf in Germany a few years earlier. Some anthropologists concluded that it was probably the remains of a "diseased Mongolian" who had died during the Napoleonic Wars. Subsequent findings revealed that the bones belonged to a far more ancient humanlike creature. Little could be concluded at the time beyond the notion that the individual, later named *Neanderthal Man*, had roamed the valley (presumably with companions) some fifty thousand or even two hundred thousand years ago, hunting and killing now-extinct mammals with spears and stone tools.

Neanderthal Man became the prototype for the fancied caveman who, grunting and slouching, clubbed his woman and dragged her off to his cave by the hair. To put it mildly, this scenario was not suggested by the evidence. We now know that Neanderthals were abundant in Europe, the Middle East, and Western Asia between 130,000 and 35,000 years ago. The Neanderthal brain, even larger than that of a modern human, was housed in a massive skull with a protruding face and heavy bony ridges over the brow. Neanderthals took care of their sick and buried their dead, often placing food and weapons, sometimes even flowers, with the bodies. These discoveries suggest thought processes associated only with humans, and we now recognize the Neanderthal as a subspecies, *Homo sapiens Neanderthalensis*.

Neanderthal watchers soon turned their attention to another early human type, many of whose bones were first unearthed in a cave in southwestern France in 1868. He became known as *Cro-Magnon Man*, after the discovery site. The Cro-Magnons made higher-tech stone tools, carved intricate necklaces and pendants, and created the extraordinary wall paintings found in caves in the Dordogne Valley of France. They may have been the first people to have full modern language capabilities. Today we accept Cro-Magnon as one of us, *Homo sapiens*.

As fossil finds of Neanderthals proliferated, anthropologists began to assume that early modern humans entered the scene after the Neanderthals, and perhaps descended from them. But in 1988 new techniques to estimate the dates of fossil finds indicated that modern humans had inhabited caves near Nazareth 100,000 years ago. It appeared that modern humans and Neanderthals may have coexisted in Europe and the Middle East for as long as 60,000 years.

These discoveries supported the growing notion that modern humans migrated from Africa to the Middle East and then spread throughout the world, while the Neanderthals entered the Middle East from Europe, perhaps escaping advancing glacial sheets. As modern humans flourished, Neanderthals became extinct. But others held fast to the notion that Neanderthals and their fellow hominids must have interbred to produce fully modern humans, who evolved simultaneously in several parts of the world.

Where does the answer lie without an abundance of additional well-defined fossils? Perhaps such bones will be unearthed eventually—or perhaps they have all turned to dust. We are left with a frustratingly small collection of bones and artifacts and a story filled with tantalizing possibilities in unfinished chapters. And yet we do have available to us what may prove to be an extraordinarily rich library of information—the genes of living humans—in which the chapters of our early history may yet prove to be preserved.

The Garden of Eden

There is near unanimous agreement that the available fossil evidence now points to a human lineage that began approximately 2.5 million years ago with *Homo habilis*, a creature that walked upright, had a brain larger than that of apes, and was the first to use stone tools. By about 1.6 million years ago these forebears had given rise to a more intelligent human ancestor, *Homo erectus*, which spread out of Africa and into the Near East, Europe, and Asia. Many have long assumed that various populations of *Homo erectus* evolved simultaneously into *Homo sapiens*, a so-called "multiregional" scenario.

There is now, however, a growing body of genetic evidence that points to another, very different picture. In this new view of the emergence of the anatomically modern *Homo sapiens*, we evolved in only one place—Africa—and quickly swept throughout the world, replacing all other humanlike creatures, including the Neanderthals. Was this the result of widespread genocide? Perhaps the explanation is more benign, since the Neanderthals coexisted with the Cro-Magnons for thousands of years. Perhaps the latter were more adept at staving off the cold during the Ice Age winters, or an influx of new diseases accompanied their arrival which decimated the Neanderthals just as the European colonists later unwittingly destroyed native populations with smallpox and measles.

The ongoing saga of the "out of Africa" hypothesis is a wonderfully instructive example of the drama of science. The practitioners of science, bound by the rules of evidence, can be at odds over the interpretation of that evidence. Despite the vaunted objectivity of scientists, human nature does not take lightly to one's cherished hypotheses succumbing to new data. In this case the data have come from the genes. The story they seemed to have told us is the story of "Eve"—but with a surprise ending.

In 1987 Rebecca Cann, Mark Stoneking, and Allan Wilson, working at the University of California at Berkeley, made a stunning announcement. They had analyzed the DNA of 147 people from five continents, and their computer told them that we could trace our ancestry back to one woman, eventually known as "Eve," who lived in sub-Saharan Africa between 140,000 and 290,000 years ago. Eve was a sensation—she soon appeared on *Newsweek*'s cover in all her imagined glory.

This startling hypothesis brings us back to our earlier discussion of molecular clocks. Earlier we noted how Allan Wilson and Vincent Sarich had found a "clock" in the blood proteins of humans and our closest relatives, the apes. Over the millions of years that these two groups had evolved in their unique ways from a common ancestor, mutations in the genes controlling the production of their blood proteins had apparently been going on at a similar slow and steady pace. Analysis of the differences in the blood proteins among chimpanzees, gorillas, and humans showed that their common ancestry reaches back a mere five million years rather than the once commonly accepted figure of 20 to 30 million years.

Sarich and others had extended their comparisons to many other organisms, such as horses, dogs, camels, and elephants, building up a web of temporal relationships among these species. And then Wilson was back, with a new time-measuring tool introduced several years earlier by Douglas Wallace, then at Stanford University, for tackling the problem of modern human origins—the mitochondrial clock.

The Mighty Mitochondria

As scientists probe the minute details of cells, they find many striking dissimilarities between bacterial cells and all other types. Among the most obvious is the presence in the latter of a few to several thousand elongated cylindrical rods, the mitochondria, whose enigmatic presence was recorded by microscopists of the nineteenth century, although their function and the details of their structure were not understood until relatively recently.

The electron microscope reveals each mitochondrion to be surrounded by two membranes, the inner one wrinkled into complex folds which further analysis shows enclose minute amounts of DNA and RNA, as well as ribosomes and hundreds of enzymes. The mitochondria are vital to the cell, for it is there that oxygen is consumed to complete the breakdown of sugars and acids and the resulting energy is captured for the cell's use.

Time-lapse cinematography of living cells catches the mitochondria in the act of constant motion, changing their shape, dividing, and even fusing with one another. They remind one of bacteria—for good reason. A most remarkable hypothesis greeted in its early form by almost universal scorn has now become accepted as a reasonable and ingenious explanation for the evolution of eukaryotic cells. It is now called the serial endosymbiosis theory, or EST. Based on speculations earlier in this century, it has been enunciated in its most complete form by Lynn Margulis, now at the University of Massachusetts at Amherst.

Put simply, it says that mitochondria, along with *plastids* (the organelles where photosynthesis takes place in plants) and *undulipodia* (whiplike bodies such as cilia and the tails of sperm) originated as free-living bacteria. These entered some archaebacterial host cell (not all at once, but at various times and places), gave up their independence, and settled in as symbiotic parts of the cell. Each lost many of the characteristics which it had as a free-living form and became streamlined, as it were, for its role as a dependent member of the cell community.

Despite the mounting structural and biochemical evidence for this stunning synthesis, it remained an intriguing but "far out" idea—until evidence from the genes themselves turned many skeptics into enthusiasts. DNA from plastids was shown to be closely related to DNA from modern photosynthetic bacteria (*cyanobacteria* or blue-green algae). This was later augmented by protein, DNA, and ribosomal RNA sequencing of mitochondria as well.

The genes appear to be telling us about events which propelled life into a great leap forward, a symbiotic partnership among cells which freed the new eukaryotes from the limitations of bacterial life and sent some on their way toward ever larger and more complex forms, including humans. We often feel separate from the rest of nature, but in each of our cells may be the descendants of ancient bacteria, moving ceaselessly about in response to the commands of the genes to wrest the life-giving energy from our food.

In giving up their independence, the mitochondria relinquished many of their genes. DNA in mitochondria is relatively sparse. In mammals, the mitochondrial DNA is about 16,500 base pairs long, less than one-hundred thousandth the length of the DNA in the nucleus. Analysis of the latter, however, clouds our picture of genetic evolution within the human species because mutations accumulate slowly in nuclear genes. Moreover, nuclear genes are passed on to the offspring from both parents and mix in every generation.

Herein lies a key to the utility of studying mitochondrial genes. Ordinarily, when a sperm and egg cell unite to form a new organism (see

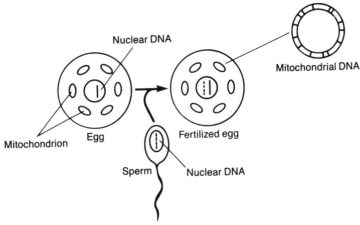

In fertilization, only the nucleus of the sperm enters the egg. The chromosomes of the sperm join those of the egg and are passed to all succeeding cells. The DNA in the mitochondria of the mother's egg is likewise passed on with her mitochondria, whereas the mitochondria of the father's sperm are left behind.

illustration) the head of the sperm, bearing only a nucleus, enters the egg, leaving the father's mitochondria behind. Only the mother's mitochondria become the mitochondria of the new generation. In our culture sons usually pass on the surname of the father to the next generation. Universally, only daughters pass on mitochondrial DNA to their progeny.

Mutations are known to occur in mitochondrial DNA and to accumulate relatively rapidly. Most do not affect function, and thus are not eliminated by natural selection. Knowing the number of mutations in populations of a known age reveals the rate at which such differences in mitochondrial DNA (mtDNA) accumulate. We can estimate the time elapsed since various lineages separated from their common ancestor by measuring the mutations. In vertebrates, mtDNA can mutate up to 10 times faster than nuclear DNA. It acts like a fast-ticking clock.

Of course, an underlying assumption must be that the ticks of the clocks in the different organisms that one wants to compare must occur at the same rate, at least on the average. Calibration is critical. It is accomplished by comparing the differences in mtDNA between two living groups and the estimated time of their divergence from a common ancestor, as determined from fossils. The accuracy of the mtDNA clock is utterly dependent on the validity of this calibration.

All About Eve

In their 1987 *Nature* paper, Cann and her co-workers calibrated their mitochondrial clock by comparing the differences within populations in New Guinea and Australia as well as the New World. They concluded that the average rate of mtDNA mutations within humans was between two and four percent per million years, a rate similar to other estimates for apes, monkeys, horses, rhinoceroses, mice, rats, birds, and fish. Given the 0.57% divergence found among existing separate human populations, they concluded that the common ancestor of all surviving mtDNA types existed some 200,000 years ago. She became known in the popular press as Eve, "an obvious and convenient term," acknowledged Wilson, although "it caused us a lot of problems."

One problem was a general misunderstanding of what was implied by the notion of one ancestral mother. Assuming a generation time of twenty-five years for humans, 8,000 generations have come and gone since this Eve is said to have existed. This suggests that a severe "bottleneck," a relatively severe population crash, must have occurred more recently so that most mitochondrial lineages were eliminated. But evidence was mounting from other sources to indicate that population bottlenecks in human history are extremely unlikely. "The garden of Eden, it turns out, was fairly crowded," quipped Craig Packer of the University of Minnesota.

However, back in 1983 John Avise of the University of Georgia had already pointed out that it was indeed theoretically possible that a diversity of mitochondrial DNA types in an ancestral population could eventually be replaced by a line of descendants from only one of those types. This does not imply that there was ever, at one time, only one woman; Eve, the "one lucky mother" as Wilson called her, could have belonged to a population of many thousands.

Along with Wilson, Linda Vigilant, also at Berkeley, and others published a much more detailed mtDNA study in 1991. This time they sampled 189 people of diverse geographic origin, including 121 native Africans. Using a comparison of mtDNA base sequences from chimps and humans to calibrate their clock, they placed the age of our common human ancestor at between 166,000 and 249,000 years ago. Eve was still in the running as the mother of us all.

The single-point-of-origin (Eve) theory implies that all other hominids, including Neanderthals and *Homo erectus* from non-African sites, became extinct without having contributed genes to modern humans. If this is the case, modern big-brained *Homo sapiens* arose in Africa as a new species and replaced other *Homo* types there as well as in Europe and Asia.

Banished from the Garden?

There had always been vocal dissenters. Prominent among them were Milford Wolpoff, professor of anthropology at the University of Michigan and Alan Thorne, head of the department of prehistory at the Institute of Advanced Studies at the Australian National University. Their multiregional scenario traced modern populations back to when humans first left Africa at least a million years ago. "Fossil remains and artifacts," they held, "represent a monumental body of evidence, and . . . a much more reliable one."

Many agree that the fossil evidence as well as that from mtDNA supports the notion of anatomically modern humans not having been in Africa before about 100,000 years ago, meaning that Eve would have lived before the origin of modern *Homo sapiens*. Fossils also show that ancestors of *Homo sapiens* were in southern parts of Eurasia nearly a million years ago. The multiregional model, as opposed to the Eve model, holds that these ancient populations underwent changes into modern people. Similar changes in other geographic regions resulted in the same kind of evolution.

Wolpoff and Thorne were adamant that it was unreasonable to assume that within 100,000 years all preexisting hunter–gatherers were replaced by humans emerging from Africa and that none of the local women mixed with these invading modern men. Reading the bones told another story: the patterns of evolution of Australasia, China, and Europe show that their earliest modern inhabitants did not share the complex of physical features that characterize Africans.

Despite these and other objections, Eve (as yet unaccompanied by Adam—but we will get to him shortly) remained a popular figure. Then, in February of 1991, the serpent moved in. In four tersely worded paragraphs, a letter published in *Science* by Alan Templeton of Washington University in St. Louis curtly dismissed Eve.

His argument was simple and derived from neither fossils nor genes. It was based on the statistics used in the computer program for turning mtDNA data into a family tree. The program had used *parsimony analysis*, which looks for a family tree based on the fewest mutations. The trouble is, according to computer expert Templeton, there are actually millions of equally parsimonious trees and no safe way to decide which is best.

Templeton had arrived at better trees with non-African roots on his first attempt. A research team led by David Maddison already had a paper in press in *Systematic Biology* that supported Templeton's criticisms. One of the authors, Maryellen Ruvuolo, explained "We're not saying that (the origin) is definitely non-African, but that you can't tell."

While the computer analysis linking modern human beings' common ancestor to a woman who lived about 200,000 years ago in Africa has come under question, nevertheless other genetic and fossil evidence still support putting Eve in Africa. There is indeed evidence that there is greater diversity in the DNA of Africans, implying that people have lived longer in Africa, because it would take more time to accumulate that variability.

Has Eve been banished from the Garden of Eden? Perhaps not, but in a special 1993 edition of *American Anthropologist*, Templeton maintained that recent studies of DNA from many contemporary populations support the theory that modern humans evolved in different geographical locations at the same time and interbred to form a single species, *Homo sapiens*.

On the other hand, "Adam was an African Pygmy," announced Gerard Lucotte of the Collège de France in 1989. The studies in his lab as well as several others use genes, in this case large strings of them that make up the human Y chromosome. This small chromosome is found only in males, whose sex-determining chromosomes are designated (because of their shape) as X and Y, whereas females are XX. In reproduction, most of the Y chromosome is passed intact from father to son. Mutations on this chromosome could lead, in theory, back to a common paternal ancestor, not necessarily Eve's mate but at least roughly a contemporary. Lucotte's interpretation of his tests of ethnic groups from around the world pointed to Adam's closest modern counterpart being the 300,000 Aka Pygmies of the Central African Republic.

Genetic analysis of our past has only just begun. Perhaps Eve (as well as Adam) is simply much older—the clock approach is correct but the calibration needs adjusting. Perhaps help will come from Kenneth Kidd of Yale University, who may have found a "clock" in tiny nuclear DNA sequences whose structure is very stable over generations. As we increase our repertoire of gene sequences useful for analysis, the odds of penetrating to our human origins turn in our favor. This may be one of the many fruits of the Human Genome Project.

The Past's Future

In the end then we return, as science always must, to the evidence. Scientists may differ in their interpretation of that evidence or in how to assign varying weights to data from disparate sources. But as scientists they always understand that doctrine must not transmute into dogma.

In particular, our saga of the search for human origins continues unabated, as does the history of our ancestors' migrations. Support for an African homeland has come from another source: Stanford University's Luigi Luca Cavalli-Sforza has mapped the worldwide distribution of hundreds of genes. Some of the maps depend on analysis of proteins, others use nuclear DNA. From this he and his colleagues have inferred the lines of descent of the world's populations as they relate to the development of languages.

Genes, people, and languages have evolved in tandem, according to Cavalli-Sforza, though a series of migrations that began in Africa, and proceeded through Asia to Europe, the New World, and the Pacific. He concludes that with its mitochondrial clock Wilson and his co-workers "have therefore confirmed our conclusions by completely independent means."

But as always we must allow the bones to speak as well. No sooner had Eve's existence been questioned than a joint American–Chinese anthropological team announced that it had unearthed two skulls from a site in China's Hubei Province. The rarity of ancient skulls makes them *ipso facto* of great interest. These two are of extraordinary interest as they may be a transition between *Homo erectus* and *Homo sapiens*—at least 350,000 years old. If so, *Homo sapiens* may have begun in China and radiated from there.

Many of the answers that we seek about our origins lie in fossils and artifacts not yet found and in more refined and extensive scrutiny of the genes. But are these parallel paths? Must they so often seem at odds? Perhaps not. The silence of the grave may yet be broken because of a new and awesome possibility.

*   *   *

What is the life span of a gene or a gene fragment? Do genes disappear with the death of the cells that house them? After all, genes are complex DNA molecules whose codes are also reflected in the cell's RNA and proteins. Surely these all must soon fall prey to the forces of decay?

The answer to this question was a resounding "No" at a packed meeting at the University of Nottingham, England, in July 1991. Richard Thomas, head of the DNA laboratory at the Natural History Museum in London, had expected a handful of experts at the workshop on the recovery of DNA from archaeological material and museum specimens. Instead, after the science section of *The New York Times* mentioned the workshop along with a fanciful "recipe" for resurrecting a dinosaur from remains of its DNA, scientists flocked to the conference along with curious members of the press.

The conferees were more than just curious. Some 40 scientific presentations revealed that the search for ancient DNA was well under way. Allan Wilson's group at Berkeley had once again been in on the ground floor. In 1984 they had succeeded in cloning DNA from preserved tissue of the quagga, a strange African beast resembling a cross between a zebra and a horse that had been extinct for over a century.

Only a year later Svante Paabo at the University of Uppsala in Sweden cloned a 34,000-base-pair piece of DNA from an Egyptian mummy who had lived more than 4,400 years ago. Then in 1988 Paabo, by then working in Wilson's laboratory, amplified mtDNA from a human brain which had been preserved in sediment in a Little Salt Spring, Florida, bog for some 7,000 years. He and his colleagues had chosen to look for segments of mtDNA known to show variability in modern human populations. They concluded that this specimen had mitochondrial characteristics previously unknown from the New World, suggesting that such remains can not only yield random bits of DNA but also give us valuable clues to questions of population history such as kinship or migration patterns.

These findings, exciting in their own right, were just the beginning. As so often happens in science, a technological advance arrives on the scene just as researchers have reached an impasse. Methods such as techniques to determine DNA sequences arrived at propitious moments to accelerate molecular biology into the central role that it plays in biology today.

Here the catalyst for a quantum leap forward was the invention of PCR, the polymerase chain reaction. PCR, a method for creating many copies of a DNA sequence from a few originals, has become an essential and invaluable tool since the late 1980s. The DNA may come from anywhere: a drop of blood, a hair—or a fragment of brain tissue preserved for thousands of years.

Genes assumed to have long since disappeared may remain in trace quantities. Since mtDNA, unlike nuclear DNA, is present in each cell in hundreds of copies, it has become a prime object of the scientists who are now searching ancient preserved organisms from mummies to mammoths.

The use of PCR has its own peculiar difficulties. It is such an exquisitely sensitive method that great care must be taken to ensure that the DNA which one has so painstakingly amplified is from the actual sample and not from a hair or even a flake of skin unwittingly contributed by the researcher. At the 1991 Nottingham conference, Brian Sykes of Oxford University, who along with Robert Hedges had been the first to extract DNA from an ancient (5500-year-old) human bone specimen described his dismay when a cow bone from the wreck of King Henry VIII's battleship yielded human DNA, as did an Anglo-Saxon horse!

The widespread availability of PCR and its proven capabilities triggered a productive synthesis among molecular biologists, paleontologists, and archaeologists in the early 90s. Studies have proliferated on the evolutionary relationships among a wide variety of organisms including bacteria, birds, mammals, and fish. The popular press has already piqued public interest in this novel approach to the past in stories such as those about DNA from "Dima," the mammoth thawed from the Siberian tundra, or the "Iceman" whose almost intact remains were yielded up by a glacier high in the Austrian Alps in 1991.

But what of the dim vistas beyond the mummified remains? In 1990 Edward Golenberg, a postdoctoral student at the University of California at Riverside, managed to isolate and analyze bits of DNA from chloroplast genes of magnolia leaves that have been preserved in the sediment of a lake for some 200 million years.

Then, in 1992, two independent teams of scientists extracted gene fragments from extinct termites which had been fortuitously preserved in amber for 25 to 30 million years, as well as from stingless bees similarly preserved for 40 million years. One of those research teams, including Raúl J. Cano and Hendrik N. Poinar from the California Polytechnic State University, reported in June 1993 that they had isolated and analyzed DNA fragments from a 120–135 million-year-old weevil preserved in amber.

Amber is ancient sap which has hardened into a plastic-like polymer, becoming almost as impermeable as glass. Organisms trapped in the sticky sap are thus preserved from decay. Analysis of the termite DNA has already challenged a popular theory that termites evolved from cockroaches. Instead, both appear to have evolved from a common ancestor.

The reality of insects trapped in amber seemed to support the flights of fancy in the popular 1993 film *Jurassic Park* (based on the 1990 novel of the same title by Michael Crichton). Crichton portrayed scientists recovering mosquitoes and flies that had drawn blood from dinosaurs; the insects subsequently alighting on fresh tree resin, becoming stuck, and finally fossilized. The dinosaur DNA isolated from these insects was used to recreate living dinosaurs. The reality, however, is that ancient DNA is always damaged and fragmented into traces of its original complexity. The notion that one could find and reorganize these tiny traces of the past and put them into living cells to develop into long-extinct organisms is not science, but science fiction.

What about the possibility of obtaining gene fragments from well-preserved ancient bone? In 1992 the analysis of DNA from the bones of a 14,000-year-old extinct saber-toothed cat buried in the famed LaBrea tar pits in Los Angeles indicated that the saber-tooths belong to the same family as modern cats and were closely related to lions and leopards.

What about our ancestors? Do traces of analyzable DNA remain in some of the fossil bones of members of our family tree? Scientists at the Los Alamos National Laboratory are now examining Neanderthal bone. Sources close to the project are reported as verifying that DNA is present, that it is damaged, and that it is from a primate. If fragments of ancient bones ever yield readable DNA, it may be possible to set the mitochondrial clock and settle the debate over our origins. There will be no difficulty in finding out if these preliminary findings prove to be fruitful. They will be headline news.

Chapter 3

DNA IN COURT

In the quiet English village of Narborough, shortly after dawn on Tuesday, November 22, 1983, the battered body of fifteen-year-old Lynda Mann was found lying in the grass next to a clump of trees. A worker from the nearby mental hospital had stumbled onto the scene just off The Black Pad, a wooded lane which meanders alongside the grounds of the hospital. Lynda Mann had been raped and strangled.

Tests on semen found on the young victim revealed that it contained proteins indicating that the murderer had type A blood. Nine months later, after blood-typing 150 local men, the police had found thirty who were type A. The others could be eliminated as potential suspects. Was one of the thirty the killer? Three years passed without another clue.

Then, on a footpath less than a mile from where Lynda had been killed, fifteen-year-old Dawn Ashworth was raped, strangled, and hastily buried beneath a pile of hay and tree branches. Again, based on semen samples from the victim, her assailant must have had type A blood. A week later a young kitchen porter at the mental hospital confessed that he had killed Dawn. He insisted that he had nothing to do with the death of Lynda Mann. Blood tests revealed that he was not type A.

*   *   *

The village of Narborough, in the English Midlands, is only six miles from the city of Leicester. At Leicester University a young researcher, Dr. Alec Jeffreys, had recently gained the attention of the news media through the application of a novel means of identification which he had developed in his laboratory. He had applied this technique, which depends on the comparison of DNA extracted from people's cells, to several highly publicized immigration and paternity suits.

The first case, in 1985, concerned a Ghanaian boy who had been born in the United States and emigrated to Ghana to be with his father. When someone claiming to be this boy came to the United Kingdom to join his mother, brother, and two sisters, immigration authorities suspected that a substitution had occurred and that the boy who showed up was either unrelated to the purported mother or was her nephew. They would not grant him residence.

At the request of the family's lawyer, Alec Jeffreys had carried out what he termed a "DNA fingerprint" analysis. The father was unavailable, but blood samples were taken from the alleged mother, her known son, and his two sisters, as well as from the boy in dispute. Jeffreys concluded, after detailed DNA comparisons, that the boy was the mother's son "beyond any reasonable doubt." The authorities dropped the case, allowing the boy to remain with the family.

Jeffreys, by now referred to by the locals as "that DNA bloke," was approached by a detective inspector from the Leicestershire constabulary and asked if he would be willing to analyze blood and semen samples to aid in the prosecution of the confessed murderer of Dawn Ashworth. They hoped to prove that he had also killed Lynda Mann. By now quite familiar with the nearby brutal slayings, Jeffreys eagerly accepted the challenge.

First, he extracted a minute amount of DNA from the semen samples taken from Lynda Mann three years earlier. It did not match the DNA in the hospital porter's blood. Next, a similar analysis was done on semen found on Dawn Ashworth. The DNA was identical to that found in the semen on Lynda Mann. The same man had committed both crimes. He was not the man in custody.

On November 21, 1986, legal and forensic history was made in Crown Court, Leicester. The confessed murderer became the first to be exonerated as a result of "DNA fingerprinting." Criminal justice had found a new and powerful weapon. In Narborough, the police reopened the hunt for the killer. A local newscaster announced: "Police are no nearer today than they were three years ago."

Not everyone on the police force believed in this newfangled DNA business. Soon they would be asked to suspend their disbelief and participate in another unprecedented effort. Chief Superintendent David Baker had convinced his superiors to go along with a campaign of voluntary blood testing of every male resident of Narborough and the two adjoining villages of Littlethorpe and Enderby.

The testing began in January 1987. It would be described graphically two years later by Joseph Wambaugh in his dramatic account of the affair, *The Blooding*. Eight months later, a total of 4,582 men who "lived, worked, or even had a recreational interest" in the area had "volunteered" (how would it have looked to have refused?) to donate blood and saliva specimens to the police. Subjects whose blood was not type A (about half of those sampled) were eliminated. A DNA profile according to Jeffreys' directions was made from the blood samples of the rest. None of them matched that of the murderer.

Later that summer, while snacking in the Clarendon Pub in Leicester, young Ian Kelly mentioned that Colin Pitchfork, his co-worker in a nearby bakery, had talked him into taking the blood test in his place. Armed with a false ID, Kelly had given blood in Pitchfork's name. When the police confronted Pitchfork, he admitted to both murders. He became the 4,583rd and last to be "blooded." His DNA pattern was a perfect match to the DNA in the semen found on the victims' bodies. Colin Pitchfork became the first person to be convicted of murder on the basis of DNA evidence. At his sentencing the judge told him, "If it wasn't for DNA you might still be at large today and other women would be in danger."

DNA FINGERPRINTS

You would have to look for one part in a million million million million million before you would find one pair with the same genetic fingerprint, and with a world population of only five billion it can be categorically said that a genetic fingerprint is individually specific and that any pattern, excepting identical twins, does not belong to anyone on the face of this planet who ever has been or ever will be. (Alec Jeffreys as quoted in *The Blooding*)

In the autumn of 1984, Dr. Jeffreys had been working on a project whose objective was to gather more information on how genes evolve. He had decided to work with the genes which function actively in skeletal muscle since he had found some interesting repetitive sequences in these human myoglobin genes. As so often happens in scientific research, he and his co-workers began by using techniques which had been developed by others as tools to solve new problems.

In order to appreciate the significance of Jeffreys' experiments, we need as usual to go somewhat beyond simple generalities and look at the scientific principles at work. You will recall that the cells of each human being contain a unique set of twenty-three pairs of chromosomes. Each pair is established at conception when the twenty-three chromosomes from the father's sperm cell mingle with the twenty-three residing in the mother's egg.

Each chromosome is a tightly coiled molecule of DNA wrapped around clusters of proteins (see Chapter 1). The DNA copies itself in the complex chromosomal dance of mitosis, also called cell division, so that each of the trillions of cells which make up the human body contains the very same 46 chromosomes—except only the sperm or egg cells, where the number is reduced to 23. Each of those 23 is one or the other of the original pair, so that this random assortment results in an offspring with a unique mix of maternal and paternal chromosomes. Sometimes, by sheer chance, the two cells which result from the first cell division after fertilization move apart and develop separately. If all goes well, identical twins result. Each has exactly the same chromosomes as the other. No other two people do.

We saw earlier that DNA is a beautifully symmetrical molecule in the shape of a double helix—the configuration one would get by twisting a flexible ladder around a cylinder. The sidepieces of the ladder are alternating units of deoxyribose sugar (S) and phosphate (P). They make a continuous strand of S-P-S-P-S, and so on.

The rungs in this ladder are pairs of nitrogenous bases. Because of their shape, adenine can pair only with thymine and guanine with cytosine. Taken together, the chromosomes are a giant twisted ladder with about three billion chemical rungs. On either side of the helix the sequence of nitrogenous bases, for example ATTCGGTA, etc., makes up the thousands of genes, each of which bears a code which directs the cell to make specific proteins.

Each gene is on a specific site, or *locus*, on the chromosome. Since the chromosomes are paired, so are the genes. A pair of genes on the same loci of a chromosome pair are referred to as *alleles*. Of course, since one chromosome in a pair is of maternal and the other is of paternal origin, the alleles may be the same or different.

While each person has a unique sequence of these nitrogenous bases spread out along his or her chromosomes, long stretches of DNA are identical among all of us. After all, we all need the same basic parts (teeth, skin, kidneys, etc.), so that much of our genetic code is shared by all humanity. Although there is one site of variation about every 1000 bases, finding unshared sequences in the code had to wait for the sophisticated analytical methods of the molecular biology revolution.

One vital spin-off of that revolution was the discovery of restriction enzymes (see Chapter 1). We now know of several hundred different varieties of these bacterial enzymes, each of which recognizes specific base sequences in DNA. When a restriction enzyme comes in contact with a sequence for which it has a particular affinity—the sequence is usually four to six bases long—the enzyme cuts the DNA molecule apart at that site. Given the size of the chromosomes, if one extracts chromosomal DNA from one human source and cuts it up with just one restriction enzyme, the result will be millions of tiny fragments.

If one were able somehow to sort out and arrange those fragments according to size, the resulting pattern would be the very same each time cells from the same person were tested. Were we to compare the pattern of cut-up DNA from two people there would be slight differences, because the base sequence would be different here and there in each DNA sample. This means that the sites that the restriction enzyme recognizes occur at different intervals here and there along the length of the chromosomes.

This procedure is known as *restriction fragment length polymorphism (RFLP) analysis*. This simply means that the restriction enzymes reduce the DNA to fragments of different sizes. But how can one distinguish a pattern out of millions of microscopic bits of DNA?

This is done by taking advantage of the fact that the DNA fragments carry a slight negative charge. Using a technique called *gel electrophoresis*, a drop of a suspension of DNA fragments is placed at one end of a thin slab of a firm gel. This is then immersed in a special solution, an electric current is applied, and the DNA moves slowly away from the negative towards the positive pole. The smaller fragments move faster than the larger ones. After this separation, which typically takes several hours, the DNA pieces have been spread out along a narrow path or "lane" in the gel.

At this point we are left with a gel bearing a line of discrete piles of DNA pieces, spread out according to size. In order to see the DNA we can simply soak the gel in a fluorescent stain that binds to the DNA and causes it to glow under ultraviolet light. This makes for a beautiful picture but one which is too compact and cluttered for the kind of discrete analysis we want to make (see illustration).

For a better look we first need to transfer the DNA pattern on the gel to a thin membrane. This procedure is called "Southern blotting," after its inventor, Edward M. Southern of Edinburgh University. The membrane is placed on the gel. Some DNA from each pile in the lane is absorbed onto the membrane. The DNA-impregnated membrane is removed, bearing the still-invisible pattern first formed in the gel.

Before the membrane is laid over it, the gel is flooded with a solution of sodium hydroxide. This "denatures" the DNA, that is, it forces apart the base pairs (AT, GC) so that the double helix ladder comes apart, leaving only the sidepieces (S-P-S-P, etc.) intact with their attached bases sticking out, no longer paired. This is now called single-stranded DNA.

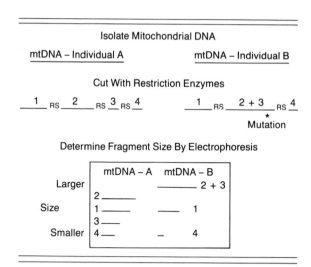

Restriction site analysis of a small portion of mtDNA is shown. Individuals A and B differ by a mutation eliminating the central restriction site (RS). This difference expressed as different sizes of fragments is determined by electrophoresis. The size of fragment 2+3 (individual B) is equal to the sum of the sizes of fragments 2 and 3 (individual A). (From Robert D. Seager, "Eve in Africa," *The American Biology Teacher* **52**(3):145, 1990. Reprinted by permission of NABT, Inc. ©1990 NABT, Inc.)

Why do this denaturing? Because now we can expose our membrane to "probes." A typical probe is a short segment of single-stranded DNA whose base sequence is complementary to another stretch of single-stranded DNA somewhere on the membrane. In other words, when the membrane is flooded with a solution containing a probe with the sequence ATTGCCGTA, it will stick to any exposed base sequence on the membrane which has the complementary sequence, TAACGGCAT. This allows one to highlight any particular segment of the DNA in the sample, depending on what probe is used.

Still, the DNA on the membrane is invisible. But we have seen to it that the probe also has either a radioactive or a fluorescent dye component attached to it. If it's the former (the type most commonly used), after washing the membrane to remove all unattached probes we can place an X-ray film over the membrane and leave them together overnight. When the film is developed, wherever a radioactive probe is attached, a black spot appears on the developed film (see illustration).

We end up with a lane on the film which appears as a series of dark bands resembling a railroad track with irregularly spaced ties. The position of each tie is a measure of the size of the polymorphic fragment. Going back to the original membrane, we can wash off the probe and apply a new one to identify a new set of fragments.

RFLP analysis of human chromosomes began in earnest in 1980 as a way of making a "map" of the human genome by identifying particular segments of DNA which were always found in the chromosomes of people with a particular genetic disease. When a probe detected a bit of DNA found in carriers of that disease but not in other members of the family, the occurrence of that particular DNA could serve as a "marker" for the disease. RFLP analysis could be used to diagnose the presence of the disease genes and trace their inheritance. This has proved to be successful for locating markers for Huntington's disease, cystic fibrosis, Duchenne muscular dystrophy, and others.

Finding an occasional abnormal gene may be sufficient for tracking the course of disease inheritance, but it is not sufficient for identifying individuals precisely. Here, at long last, we come back to Alec Jeffreys. He knew that in 1980 Arlene Wyman and Ray White, then working at the University of Massachusetts at Worcester, had found a small region in human DNA that exhibits an unusually high number of variations in base sequence. A few other hypervariable areas had subsequently been found in other laboratories.

These regions consist of side-by-side repeats of a short base sequence, or "minisatellite." Jeffreys was working with one of these short minisatellite regions, which he had earlier found in the human myoglobin gene. He discovered, much to his amazement, that when he used his minisatellite as a probe, rather than just pairing with one or two DNA

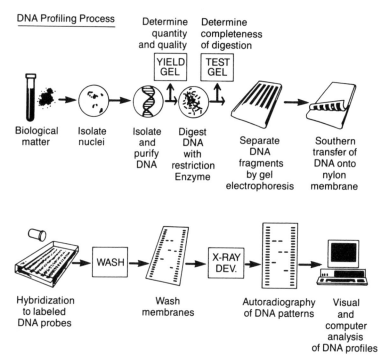

DNA Profiling Process   Determine   Determine
                         quantity   completeness
                         and quality   of digestion

                         YIELD   TEST
                         GEL   GEL

Biological   Isolate   Isolate   Digest   Separate   Southern
matter   nuclei   and   DNA   DNA   transfer of
                         purify   with   fragments   DNA onto
                         DNA   restriction by gel   nylon
                                    Enzyme   electrophoresis membrane

                    WASH   X-RAY
                                      DEV.

Hybridization   Wash   Autoradiography   Visual
to labeled   membranes   of DNA patterns   and
DNA probes   computer
                                                              analysis
                                                              of DNA profiles

This is how DNA profiles are prepared. When the DNA fragments are separated
by gel electrophoresis, samples of each fragment are transferred to a membrane
(the so-called Southern blot). Probes are applied, and the probes stick to those
DNA fragments which have bases complementary to those on the probes. In this
illustration, the probes are radioactive, so that exposure of the membrane to an
X-ray film results in a pattern of bands, which can be scanned and analyzed by
computer. (From Pamela Knight, "Biosleuthing with DNA Identification," *Bio/
Technology*, June 1990, pp. 506–507. Reprinted by permission of Nature Publishing Co. ©1990 Nature Publishing Co.)

spots on the gel, the developed X-ray film had a whole series of bands resembling the now-familiar bar codes used to label grocery items.

It turned out that he had discovered a probe which zeroed in on a whole complex of areas in human DNA which could vary enormously between individuals, even between parent and child. Later research by Jeffreys and others revealed that these tandem repeat sequences can range in number at any locus in the DNA molecule from four to hundreds of repeats. It did not take long for Jeffreys to realize the practical implications for using this means of distinguishing DNA from different sources.

As he put it succinctly in his March 1985 article in *Science*: "We anticipate that these DNA 'fingerprints' will . . . provide a powerful method for paternity and maternity testing (and) can be used in forensic applications."

> We see it (DNA fingerprinting) as probably the most significant thing
> for the century. (John W. Hicks, deputy assistant director of the FBI, in
> *The New York Times*, November 2, 1988)

Within a year of Jeffreys' discovery, scientists at Lifecodes Corporation, Valhalla, New York, had developed a two-probe system, each probe attached to a different highly variable region at a specific locus. As a result only one or two DNA fragments were detected as dark bands with each probe for each individual. Since the probability of two individuals having exactly the same set of alleles at this hypervariable locus by chance is minuscule, this is a highly sensitive technique, especially when the same DNA is exposed to a second, third, or fourth specific locus probe.

Currently in the United States single-locus probes are preferred

$$\longrightarrow$$

DNA "fingerprints" are a record of variations in a person's DNA. In a particular chromosomal region, for example, a DNA sequence (denoted by an arrow in the sketch at the top) may be repeated 3, 4, or 5 times, or appear only once, on chromosomes from different people. When the DNA from these regions is cut with a particular restriction enzyme (the cutting site is indicated by the vertical bars) the fragments produced differ in length because of this variable number of tandem repeats. After gel electrophoresis, the resulting pattern shows bands indicating the size of the fragments, ranging from the largest at the top to the

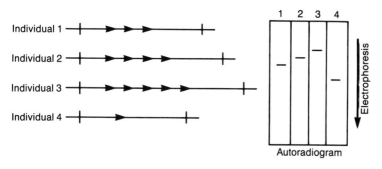

smallest at the bottom. As an illustration, gel patterns are shown for three samples probed with four different probes. Sample E is an evidence sample, while samples S1 and S2 were taken from possible suspects. For each probe, the DNA yields one band from the paternally derived chromosome and one from the maternally derived chromosome. Two bands are typically seen, although there may be only one band if the two copies of the chromosomes have fragments of the same size. Sample S2 matches the evidence sample E at the four loci considered, whereas sample S1 matches it only once. Suspect S2 would be said to be included in the set of people who might have given rise to the evidence sample, and suspect S1 would be definitely excluded as the source of the sample. (From Daniel J. Kevles and Leroy Hood, *The Code of Codes: Scientific and Social Issues in the Human Genome Project*, Harvard University Press, 1992. Reprinted by permission of Harvard University Press. ©1992 Daniel J. Kevles and Leroy Hood.)

because they generate fewer bands and are easier to interpret. Their discriminating power is greatly enhanced by using several—usually four—different probes for one analysis (see illustration). In 1988 the Cetus Corporation in Emeryville, California, announced that it had applied its revolutionary process, the polymerase chain reaction (PCR), to copy DNA samples containing less than one billionth of a gram of DNA. Now virtually any amount of DNA no matter how minute (in theory, less than the amount in one cell) could be transformed into millions of copies in less than a day in the laboratory. Cetus has developed a DNA typing kit based on PCR amplification followed by immersion of a test strip containing probes into the amplified DNA.

Regardless of the method used—and variations of these three main themes have since been developed—it is convenient to refer to such DNA analyses as *DNA typing*. The use of DNA typing spread like wildfire through a broad range of disciplines—diagnostic medicine, animal and plant evolution research, family relationship analysis, and wildlife and human forensic science. Our particular interest here is in the forensic uses. The first conviction in the United States based on DNA typing occurred in Florida in a November 1987 rape trial. Hundreds more cases soon followed including homicides, rapes, robberies, and hit-and-run accidents.

The FBI heralded DNA typing with undisguised enthusiasm. Director William S. Sessions had reason to be optimistic, given the prevailing attitude toward the technology. As one defense attorney put it: "If they print your guy with this stuff you're dead. You can't combat it. There is no defense to it."

The early enthusiasm among law enforcement officials was understandable. There were and still are well over one million violent crimes reported annually in the United States. These include approximately 90,000 rapes, 20,000 homicides, and 700,000 aggravated assaults. Of the forcible rapes, reported and unreported, over 60 percent remain unsolved, as are more than 25 percent of the homicides. Each year approximately 2000 dead children cannot be identified.

The prospect of a powerful new tool to resolve many such legal cases was considered nothing short of miraculous. A semen stain, a drop of blood, or even a strand of hair might be enough to pinpoint a perpetrator or exonerate an innocent suspect. Wrapped as they are in highly technical jargon, the telltale black spots on the X-ray film at first seemed invincible.

But even as the FBI was beginning a major effort to get DNA typing into the hands of local and state law enforcement agencies, the House Judiciary Subcommittee on Civil and Constitutional Rights held a hearing in Washington, D.C., at which witnesses voiced serious concerns about the lack of laboratory controls and accepted standards, a deficiency that could result in the conviction of innocent citizens. A storm was brewing which is still swirling through the academic journals, both legal and scientific, and most importantly through the courts.

In order to understand the nature, evolution and implications of the controversy, let's trace its principal events over the relatively few years that DNA typing has been available.

> It is impossible under the scientific principles, technology and procedures of DNA fingerprinting (outside of an identical twin) to get a 'false positive.' (Joseph Harris, a county judge in Albany, New York, in *The New Republic*, April 3, 1989)

Until early 1989, California Attorney General John Van de Kamp had been opposed to using DNA fingerprinting in criminal trials, warning about getting "mesmerized with DNA's potential" and rushing heedlessly into court. Then, after a year of deliberation during which California's Association of Crime Lab Directors evaluated the private DNA labs and endorsed their procedures, he announced that "DNA is ready to go to court, and to win."

By that time criminal trials in more than a dozen states had admitted DNA evidence, and its use had been upheld by a Florida appeals court in October 1989. However, DNA typing had not yet been put to the demanding Frye test that California courts and many other states use to determine the admissibility of new scientific techniques.

The Frye test is named after a 1923 ruling by the U.S. Court of Appeals for the District of Columbia Circuit. It requires a new scientific technique to have won general acceptance in the relevant scientific community before it can be used in court. This test would prove to be a bone of contention in the years to follow.

Van de Kamp had other plans for genetic fingerprinting, however. On April 7, 1989, he urged the creation of a network of regional labs run by state and local law enforcement agencies. Then he proposed that the state establish a computerized genetic database for everyone convicted of a violent crime. Under this proposal such individuals would be obliged to give two blood specimens and a saliva sample upon entering prison.

DNA analyses of these would be fed into a central computer, allowing police to search for a match in much the same way that ordinary fingerprints are compared. In addition to the expected annual 8,250 specimens, an additional 8,000 tests would be done on frozen blood samples taken from sex offenders since 1983.

Similar plans were under consideration in Washington State, Colorado, and Virginia. Moreover, the FBI had begun work on a computer system that could ultimately link all states together. Despite the fact that Deputy Assistant Director Hicks of the FBI had announced the bureau's plans to develop a working group to look at controls on access to the planned files, some observers, including civil liberties groups, had serious misgivings.

They argued, for example, that such records in centralized databases could afford the temptation to engage in wholesale computer-based searches for the owner of a hair or drop of blood left at the scene of a crime. Such a search without linking the individuals to the crime scene could represent a violation of due process and civil liberties. Also, databases could conceivably be scanned by whatever agency or individual gained access to the system.

One ready answer to such misgivings was that whatever invasion of privacy might be inflicted on sex offenders or other violent criminals would be far less than the invasion of privacy suffered by their victims. That reaction, while appealing in one sense, did not answer the legal, constitutional, or ethical objections.

As far as the question of gathering a biological sample from suspects is concerned, their civil rights must be safeguarded in that instance as well. Before arrest, biological samples may be taken from a suspect but only pursuant to a search warrant, based on probable cause. After arrest, the prosecution may obtain a warrant or court order directing the defendant to allow such samples to be taken. If the defendant refuses, such refusal can be introduced at the trial as circumstantial evidence of guilt.

As of late 1992 the State of Virginia had collected 50,000 blood samples from serious offenders, and 2,000 samples were being added each month. A federal court ruled on March 4, 1991, that the state had a "special law enforcement need" for the DNA samples, calling the intrusion of taking a blood sample a minor one. Nineteen other states had passed legislation allowing police to take biological samples from convicts. It is estimated that the number of states with DNA data banks will increase to 30 by 1995.

According to Paul Ferrara, director of Virginia's Division of Forensic Sciences, a search of the state's data bank for a DNA profile that matched one from a crime scene could bring up "near matches" as a "list of suspects." At a September 1991 meeting sponsored by the National Institutes of Health concerning forensic data banks, participants warned of the danger of such a procedure becoming an exercise in "rounding up the usual suspects."

Another long-term privacy concern has arisen over the possibility that certain genetic markers in a DNA profile might someday be shown to be linked to a disease or an undesirable trait. Perhaps certain DNA fragments could be associated with behavioral traits. As the efforts of the Human Genome Project continues to expand the amount of information that we can read from a DNA analysis, concerns over the confidentiality of stored biological specimens will persist.

These concerns are not limited to the United States. In early 1992, "Liberty," a British civil rights group, challenged the legality of London's Metropolitan Police DNA database. In Britain, the Home Office Forensic Office has set up a central database similar to that now planned by the FBI for general use in 1993.

> DNA typing "can constitute the single greatest advance in the 'search for truth' and the goal of convicting the guilty and acquitting the innocent, since the advent of cross-examination." (A New York State trial judge, in *People* v. *Wesley*, 1988)

> It is my contention that DNA forensics sorely lacks adequate guidelines for the *interpretation* of results—both in molecular biology and in population genetics. (Eric S. Lander, in *Nature*, vol. 339, 15 June 1989)

In the first two years after DNA typing had traveled from England to the United States it figured in dozens of criminal trials. Investigators would compare the suspect's pattern found after applying certain DNA probes to the Southern blotting membrane with the pattern derived from a body fluid stain or other biological evidence found at the scene of the crime.

If the patterns matched, scientists from one of the three private DNA typing laboratories in this country would be asked to testify in court about the chances that such a match could be coincidental. The estimates would all set the odds of an accidental match as extraordinarily slim. The numbers given would reflect the particular method used in the analysis, and would typically run from one in 30 billion to one in tens of thousands.

In February 1989, results released at a forensics conference in Orange County, Florida took some of the luster from what appeared to be an almost invincible weapon against crime. The sheriff's department had sent 50 blood and semen samples taken from about 20 people to each of the three laboratories, asking for a determination as to which specimens came from the same people.

Taken together, the laboratory results were 98 percent accurate. Cellmark Diagnostics and Forensic Science Associates had each erred on one match. They blamed the mistakes on laboratory errors. Even assuming that to be the case, the 2 percent error rate focused renewed attention on the question of exactly what level of credence one could place in DNA evidence.

Only three months later, in May 1989, a murder trial opened in the New York City borough of the Bronx which would deal a much more serious blow to the public image of forensic DNA testing. The analysis of bloodstains presented as evidence for the prosecution opened up a long, contentious struggle which ended with a ruling that was hailed as a victory by both the prosecution and the defense. The judge called for all attorneys in previous DNA cases to search for indications that appeals might be warranted. What had happened?

THE CASTRO CASE

The blood-soaked body of Vilma Ponce, 20 years old and seven months pregnant, was found by police in her Bronx apartment on

February 5, 1987. She had been stabbed 61 times. Her two-year-old daughter Natasha lay in an adjoining room, also dead from multiple stab wounds. A neighbor, Joseph Castro, had been seen acting suspiciously by several witnesses on the day of the murders. Jeffrey Otero, husband and father of the victims, not yet aware that the slayings had occurred, had seen Castro leaving the alleyway next to the apartment and noticed that he had blood on his face, left wrist, and sneakers.

The suspect was questioned and released on February 7. The police did not yet have enough evidence to hold him. However, during the questioning, one of the detectives spotted what appeared to be a drop of dried blood on Castro's watch. Castro agreed to let police retain the watch for "inspection." On March 5, 1987, the 30-year-old Hispanic Joseph Castro was arrested and charged with two counts of murder.

Lifecodes Corporation, developers of the DNA-Print Identification Test, had been asked to compare DNA from the blood on the watch with that extracted from the blood samples taken from the victims during autopsy. They did so and reported on July 22, 1987, that the blood on the watch matched that of Vilma Ponce. They further stated that the frequency of this DNA pattern was one in 100 million within the U.S. Hispanic population.

The wheels of justice often grind slowly. The case did not come to court until February 1989. Uneasy about the still-novel DNA evidence, the defense attorney elected to have it challenged in a pretrial hearing. This was a "Frye hearing," a forum in which the acceptability of novel scientific evidence is put to the Frye test. The Frye test has three essential elements. The evidence must be based on a valid theory, a valid technique applying that theory, and on the proper carrying out of that technique.

What determines validity? According to the 1923 decision on which the Frye guidelines are based, "The thing from which the deduction is made must be sufficiently established to have gained general acceptance in the particular field in which it belongs." The highly publicized pretrial hearing in *People* v. *Castro* became more than a judgment of this particular case. It put the whole of forensic DNA typing on the witness stand.

New York attorneys Barry Scheck and Peter Neufeld joined the defense team. They were fresh from a meeting on forensic DNA typing at the Banbury Center at the famed Cold Spring Harbor Laboratory on Long

Island, New York. There, molecular biologists, prosecution and defense lawyers, FBI representatives and other interested parties had wrangled over the issue of problems which could arise with the techniques used to perform and interpret DNA typing.

Also present at the Banbury meeting was Eric Lander, a talented mathematician and former Rhodes scholar who is now the director of the Center for Genome Research at MIT. Lander was disturbed by what he felt were serious deficiencies in the conclusions drawn from DNA data and described in the presentation by Lifecodes and Cellmark Diagnostics. A few weeks later he agreed to appear as a witness in the Castro case (declining the usual $1,000-a-day expert witness fee, believing it would constitute a conflict of interest).

The pretrial Frye hearing, held in the Bronx County Courthouse—known locally and to the readers of the best selling novel by Tom Wolfe, *The Bonfire of the Vanities*, as the Fortress—lasted an unprecedented 12 weeks and amassed 5,000 pages of testimony. When it was over, Lander would write in an exquisitely detailed critique in the June 15th edition of *Nature* that

> Although DNA fingerprinting clearly offers tremendous potential as a forensic tool, the rush to court had obscured two critical points: First, DNA fingerprinting is far more technically demanding than DNA diagnostics; and second, the scientific community has not yet agreed on standards that ensure the reliability of the evidence.

Scientists from Lifecodes had probed the Southern blots done on the blood samples with three probes, along with an additional probe to determine the sex. As we noted before, the Lifecodes report to the district attorney stated that the DNA from the watch and the mother's blood were a match, with the estimated frequency of the DNA pattern being about 1 in 100 million in the Hispanic population. No difficulties or ambiguities were indicated.

However, the pretrial testimony revealed a more complex story. In the case of the sex determination, the lane on the gel marked "control," that is, the lane which should contain a sample from a male as an internal check to verify the accuracy of the procedure, showed no bands at all. The mother, daughter, and watch lanes also had showed no bands. Who was the control?

At first the control DNA was said to have been from female cells in a cell laboratory culture. Two weeks later the technician who actually had performed the test claimed that the control had come from a male scientist. Then a company representative explained that this male had a rare genetic abnormality in his Y chromosome which had altered the sequence normally detected by the probe. Later, the laboratory director reported that the control blood had actually been drawn from a female laboratory technician. Apparently, Lifecodes had not specifically recorded the identity of the control blood donor.

To muddy the waters even further, the results of using the probe for one of the three DNA loci examined showed five bands in the watch lane and only two in the mother's. Michael Baird of Lifecodes asserted that the two additional bands were "contaminants." Howard Cooke of the Medical Research Council in Edinburgh, who had discovered the locus in question and supplied the probe to Lifecodes, testified that in the absence of any experiment to explain the extra bands, the defendant would have to be excluded.

Lifecodes' laboratory records indicated that three bands had been found in the DNA pattern of the daughter, in the same position as the three bands in those of the mother's blood and that on the watch. In fact, the witnesses for both the prosecution and the defense testified that there was only one band. Moreover, the forensic report listed only two of the bands!

The criteria for deciding if bands actually were to be considered as matching was that, according to Lifecodes' own written standards, their measured sizes should be within 1.8 percent of one another. The Lifecodes laboratory records showed that the bands at two of the loci fell outside that margin, and yet they were called a match. Baird admitted that the comparisons had been made by eye.

And what of the 100-million-to-one likelihood that the blood on the defendant's watch was not really that of the slain mother but someone else's who happened to have her same DNA print? This was yet another point of contention in *People* v. *Castro* and has remained at the center of DNA typing controversies ever since. It revolves around the complex question of how to calculate the frequency of alleles in a heterogeneous population.

Avoiding the numerical details, the problem goes something like this. What are the odds that a DNA pattern could match another simply by chance? That's something like asking the following question: What are the chances that on a given day you will see more than one seven-foot-tall, red-headed, bearded male smoking a cigar while driving a pink convertible down the main street of town? Pretty slim, you might answer, and ordinarily you'd be right.

The actual odds against seeing the same phenomenon more than once might be expressed by multiplying the odds of each category, that is, multiplying the expected frequency of exceptionally tall males times that of red-headed men times that of pink convertibles, etc. One would have to know that the incidence of each category was independent of all the other categories. If it turned out that your hometown required men to grow beards and was a haven for outlandish convertible collectors, the odds would be considerably less than one might ordinarily expect.

In *Castro*, what attention had been given to calculating frequencies of the alleles in a heterogeneous population? Lifecodes had used a database of U.S. Hispanics. Their data, as analyzed by the scientists at the hearings, contradicted Lifecodes' conclusions. It turned out that the odds of two individuals in the Hispanic population studied sharing the same alleles was considerably better than Lifecodes had stated. In fact, this error as well as inconsistencies in the way in which band matches were analyzed accounted for at least an 8,000-fold error. Lifecodes' honesty or integrity was not at issue. The crux of the matter was quality control and procedural accuracy.

The scientists testifying on both sides of the case had heard enough. On the morning of May 11, 1989, in a borrowed office in Manhattan, the scientists (minus Michael Baird, who had a previous engagement) gathered, with no lawyers present, and once again reviewed the evidence. They agreed unanimously that it was seriously flawed. Their consensus statement declared:

> The DNA data in this case are not scientifically reliable enough to support the assertion that the samples . . . do or do not match. If these data were submitted to a peer-reviewed journal in support of a conclusion, they would not be accepted. Further experimentation would be required.

At first, the prosecution indicated that it would withdraw the DNA evidence. The district attorney decided to go forward, and the prosecution mounted an ill-fated rebuttal. Faced with the unusual spectacle of scientists on both sides agreeing after a heart-to-heart talk, Acting Justice Gerald Sheindlin of State Supreme Court ruled on August 14, 1989, that the sophisticated genetic tests that had linked the murder suspect to a victim were not scientifically reliable. Both the prosecution and the defense claimed a victory. Even Lifecodes, whose work had been roundly criticized, hailed the decision.

Actually, a careful reading of the ruling would seem to permit at least a Pyrrhic victory for DNA evidence. Justice Sheindlin, while judging the particular DNA analysis under question to be unreliable, also ruled that DNA typing identification techniques are generally reliable and admissible as evidence. Lifecodes' senior vice-president, John Winkler, said he was "delighted that the judge recognized that using DNA for forensic testing is valid." He added that Lifecodes agreed to all of the judge's recommendations for improving the care with which such testing is done.

The two defense attorneys, Barry Scheck and Peter Neufeld, formed a committee with the National Association of Criminal Defense Lawyers to serve as a resource for attorneys who might want advice in re-opening cases involving Lifecodes. They felt that the ruling had not gone far enough. The judge, in their opinion, should have ruled that all then-current DNA testing procedures failed to meet the Frye legal standard. Scheck called for a moratorium on the forensic use of DNA analysis until proper guidelines were in place.

According to Eric Lander, "Clinical laboratories have to meet higher requirements to be allowed to diagnose a case of strep throat than a forensic laboratory has to meet to put a defendant on death row."

As for Joseph Castro, he pled guilty to two counts of second degree murder. He admitted that the blood had gotten onto his watch while he was stabbing one of his victims.

By this time the scientific, legal, and forensic communities were clamoring for a detailed examination of DNA typing, particularly its reliability and validity. They turned to the National Research Council (NRC) of the National Academy of Sciences (NAS). The NAS is a private, nonprofit society of distinguished scholars. It was founded during the Civil War by Abraham Lincoln to advise the government on technical matters. Woodrow Wilson added the NRC during the first world war as the operating arm of the Academy.

The Academy had, in fact, attempted to initiate such a study in the summer of 1988, but could not convince the National Institute of Justice to come up with $300,000 to fund it. This time around, however, NAS received support from a number of federal agencies including the National Institute of Justice, the FBI, the National Institutes of Health's National Center for Human Genome Research, and one private agency, the Alfred P. Sloan Foundation. The Committee on DNA Forensic Technology in Forensic Science was founded and its first meeting was held in January 1990. Eric Lander was among the 12 members, all highly respected and prominent figures from the worlds of science, medicine, ethics, and law.

The Committee report was finally made public in April 1992. According to population geneticist Richard Lewontin of Harvard in his account in *The New York Review of Books* of May 28, 1992, the committee had been asked "to produce a definitive report and recommendations. They have now done so, adding greatly to the general confusion." We shall see why presently.

During the more than two years between the initiation of deliberations and the NAS report, DNA forensics continued to make the headlines. Even earlier, on August 14, 1989, the very day that the *Castro* decision had ruled out the use of much of the DNA evidence, DNA typing had freed Gary Dotson.

Dotson had been sentenced to 25–50 years in 1979 for raping Cathleen Crowell Webb. In 1985, filled with remorse, she admitted that she had wrongly accused Dotson. His sentence was commuted to the six years he had already served. In 1988 Forensic Science Associates of Richmond, California, concluded that Dotson could not have been the source of the semen found in Webb's underwear.

The clothing had been stored in a police freezer for 10 years, and yet sufficient DNA had been extracted from the sample to compare it with DNA from Dotson's blood. He was granted a new trial, and the state, faced with the DNA evidence, declined to prosecute.

In Virginia, the first murder conviction in the nation based on the use of DNA typing was upheld by the state Supreme Court in September 1989. The Court affirmed a trial court's finding in *Spencer* v. *Commonwealth* that Timothy Wilson Spencer had been reliably linked to the rape and slaying of two women. The state Supreme Court decision held that the testing techniques were reliable and generally accepted in the scientific community and that the specific test used had been done in a reliable manner.

In November 1989 a Delaware trial court held in *State* v. *Pennell* that DNA evidence was admissible, but refused to admit the probability statistics. The defense claimed that the assessment of the probability of declaring a match was overstated. In its ruling the court expressed its concern that the estimates of the distribution of alleles in a population could vary depending on who was doing the measuring and decided that the evidence presented had not supported the probability calculation.

In Minnesota, Thomas Schwartz had been charged with murder in a stabbing death. Prosecutors claimed that DNA tests done on his blood-stained clothing linked Schwartz to the murder. His lawyers objected to the introduction of the DNA evidence. The Minnesota Supreme Court ruled in November 1989 that although DNA typing had gained general acceptance in the scientific community, Cellmark Diagnostics of Germantown, Maryland, who had done the testing in this case, had deficiencies in their quality control procedures. Moreover, they had refused to supply sufficient information to the defense so that it could adequately evaluate the tests.

James E. Starrs, Professor of Law at the National Law Center of George Washington University in Washington, D.C., called the ruling "prudent, but possibly overcautious." Barry Scheck, defense attorney in the *Castro* case, lauded the decision. He regarded it as a step forward that defense lawyers were learning how to challenge these technical tests.

It is obvious from these few examples that attitudes toward the use of DNA, in criminal trials at least, had evolved by the end of the 1980s from unquestioning awe to a more sophisticated analysis of individual circumstances. What was often at issue was the problem of evaluating the statistical assumptions that the companies doing the analyses made to come up with their odds that a particular DNA match was valid.

Some other investigative uses of DNA continued unabated. Annually, almost 285,000 paternity suits are filed nationwide, of which 60,000 are disputed and require paternity tests. There are, in fact, tests for disputed paternity that are much more sophisticated than simple blood-type analysis. Even if a man is found to have a blood type compatible with being the father, this would only mean that he may be the father, and not that he definitely is.

Over the last ten years, much more specific comparisons of blood proteins and enzymes have been used. If a man is not excluded by these tests, the probability that he is the father can often be shown to be as high as several thousand to one. Careful DNA typing might increase the odds considerably. John Huss, vice president of Cellmark Diagnostics, reported in July 1989 that his company had done more than 2,000 DNA paternity tests over the previous 14 months, and not one case had gone to trial.

> We are satisfied that DNA testing performed by competent, properly trained personnel can provide reliable results that meet the needs of the criminal justice system and satisfy the requirements of the courts. (John W. Hicks, Assistant Director, Laboratory Division, FBI, in *The New York Times*, February 2, 1990)

> Major problems have been demonstrated with the methods used (which have not been standardized) by the country's two major private testing laboratories, neither of which have published their methodologies in peer-reviewed journals. Prosecutors should accordingly seriously consider a voluntary moratorium on the use of DNA fingerprinting evidence to identify individuals as suspects, at least until the National Academy of Sciences reports on the scientific protocols that should be used to perform and validate DNA fingerprinting. (George J. Annas, "DNA Fingerprinting in the Twilight Zone," in *Hastings Center Report*, March/April 1990)

Eight months after the National Research Council had begun its own deliberations (which would extend seven months beyond their intended 14-month duration), the Office of Technology Assessment (OTA) weighed in with a 204-page report on forensic DNA testing. The OTA is a nonpartisan study group reporting to the United States Congress. Its basic function is to help legislators to anticipate and plan for the results of technological changes and to examine the ways in which technology affects peoples' lives.

Massachusetts Senator Edward M. Kennedy had requested the study as chairman of the Labor and Human Resources Committee. The advisory panel that had prepared the 204-page report, *Genetic Witness: Forensic Uses of DNA Tests*, was chaired by geneticist C. Thomas Caskey of Baylor College of Medicine in Houston.

The panel agreed that DNA tests are valid, as is the scientific basis of those tests. When properly performed, DNA profiles *per se* are reliable; that is, they can be reproduced within a laboratory, or among laboratories, by different examiners. However, serious questions remain about how to ensure the reliability of any single test result (see illustration).

In other words, they called for what had been the focus of concern in a number of court cases, a few of which we have cited earlier: the need to establish standards to determine what constitutes a properly performed and correctly interpreted test. They recommended that the government or professional groups move immediately to set technical and procedural standards that all could agree on.

By the time the OTA document was released on August 5, 1990, DNA profiling had been used in investigations in 49 states and the District of Columbia and by the military. Even with the increased scrutiny of the DNA procedures and interpretations, their continued use seemed to be assured. Proponents were quick to point out that in 37 percent of the over 500 rape and homicide investigations done by the FBI by January 1990, the primary suspect had been excluded, pointing to DNA profiling as a unique opportunity to protect the innocent as well as to convict the guilty.

Meanwhile, the European DNA Profiling Group (EDNAP) had launched a study of the 10 leading European laboratories producing DNA profiles for forensic cases. EDNAP had found a "considerable amount of variation" in the laboratories' results, despite the fact that the laboratories had all worked on the same samples, restriction enzymes, and probes. Cellmark Diagnostics, one of the world's leading laboratories for DNA profiling, announced that it was sending a questionnaire to 50 labs all over the world to gauge their reaction to Cellmark's offer to doublecheck the competing labs' profiles. The service would be confidential; only the lab concerned would be told if Cellmark questioned the results.

> Critics of DNA testing toss around terms like 'degradation' and 'contamination', leaving the impression that lab technicians drop samples on the floor. . . . They call for a moratorium on DNA evidence. But isn't that simply a desperate attempt to neutralize a technology they can't defeat in court? (Michael Baird, director of the paternity and forensic laboratories of Lifecodes Corporation, in *ABA Journal*, September 1990)

Magistrate James E. Carr of Federal District Court in Toledo, Ohio, had convened a six-week hearing during which 13 of the nation's leading molecular biologists and population geneticists studied every aspect of DNA forensics. The FBI, which had called it a "scorched earth review," was criticized by the magistrate in his ruling for "the remarkably poor quality" of its work and "infidelity to important scientific principles."

However, the October 1990 ruling was hailed as a victory for genetic profiling because Magistrate Carr had confirmed that the technique is admissible as evidence in a criminal trial. In a particularly controversial decision, he held that experts need not necessarily agree on novel scientific evidence in order to find that a practice is generally accepted in the scientific community. Peter Neufeld, who appeared as a defense lawyer in the case (*United States* v. *Yee*) remarked, "I think he's wrong on the law . . . you don't need unanimity, but you do need consensus."

Richard Lewontin of Harvard once again entered the fray, along with Washington University's Daniel Hartl—and with a vengeance. At issue were the astronomical figures so often cited in court of the minuscule probability that two DNA samples could match by chance. In a detailed six-page article in the December 23, 1991, *Science* they cited figures such as 1 in 738 trillion or 1 in 450 million as "terribly misleading" and "unjustifiable." They argued that until data for determining the genetic variability among ethnic groups could be gathered, which could take up

-->

This diagram illustrates a possible source of error in comparing DNA profiles. Two samples of DNA may yield patterns that are in different locations but shifted relative to one another. There are two possible explanations: the DNAs are actually from different sources, or the DNAs are from the same source but one

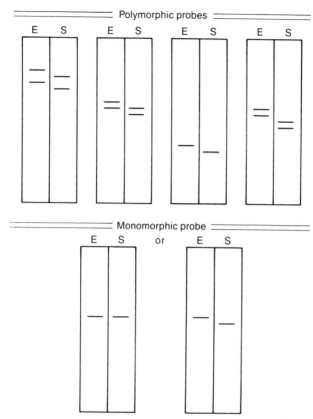

sample has migrated faster for technical reasons (such as a difference in salt concentration). The second alternative is called *bandshifting*. To distinguish these possibilities, one can use a monomorphic probe, one that detects a fragment of the same length in all samples. If this constant-length fragment is in the same place in both samples, one can conclude that there has been no bandshifting and the samples must be different. If the constant-length fragments are not in the same place, then bandshifting has occurred. One must then attempt to correct quantitatively for the extent of bandshifting. (From Daniel J. Kevles and Leroy Hood, *The Code of Codes: Scientific and Social Issues in the Human Genome Project*, Harvard University Press, 1992. Reprinted by permission of Harvard University Press. ©1992 Daniel J. Kevles and Leroy Hood.)

to 10 or 15 years, such probability statements should not be allowed in court.

The crux of their argument was simply that the current estimates of odds were based on the assumption that such groups as blacks, Caucasians, and Hispanics were each quite genetically homogeneous. This, they contended, ignored all of the evidence that each of those groups is actually made up of numerous subpopulations. Moreover, each of those subgroups is genetically diverse. For example, they called the designation "Hispanic" a "biological hodgepodge" made up of people of Mexican, Puerto Rican, Cuban, and Spanish descent, among others.

In one sense, one could argue that this was a tempest in a teapot, or perhaps in a test tube. After all, one might think that odds of 1 in 500,000 should be just as impressive to a jury as 1 in 50 million. In fact that was the position taken by Ranajit Chakraborty of the Center for Demographic and Population Genetics at the University of Texas and by Kenneth Kidd of the Department of Genetics of the Yale University School of Medicine. They co-authored a rebuttal of the Lewontin and Hartl paper in the very same issue of *Science*.

What might seem to some as merely an academic debate over statistics turned out to be seen by proponents of DNA typing as a serious threat to confidence in the technology, enough to persuade judges to deny admissibility of DNA evidence and derail the prosecution. Some accused Lewontin and Hartl of trying not merely to improve the methodology, but to keep it out of court entirely.

The authors vigorously denied any such motivations. They stressed, in fact, that "appropriately carried out and correctly interpreted . . . DNA typing is possibly the most powerful innovation in forensics since the development of fingerprinting." Nonetheless, the publication of their *Science* paper was accompanied by what appeared to be quite unusual happenings.

Unknown to the authors, copies of a prepublication manuscript of their paper were circulated before the International Congress of Human Genetics, held in Washington, D.C. in October 1991. Several attendees approached a senior editor of *Science* at the Congress and lobbied against acceptance of the paper. *Science* editor Daniel Koshland commissioned the rebuttal by Kidd and Chakraborty in order, as he put it, "to be fair."

According to an account in the August 7, 1992, *Science*, Daniel Hartl received a call in early October from James Wooley, an Assistant U.S. Attorney in the Department of Justice's Organized Crime Strike Force Division in Cleveland, Ohio. Wooley urged Hartl to reconsider publishing the paper. Hartl protested vigorously what he described as a "chilling" conversation in which he had felt "intimidated." Wooley denied any intimidation.

The Lancet, a highly respected British medical journal, in chronicling the acrimonious debate, predicted in January 1992 that once the dust had settled, the upcoming National Academy of Sciences report on DNA profiling "will be accepted as definitive guidance." As it soon turned out, that would depend on whether or not one preferred to regard an eight-ounce glass containing four ounces of water as half full or half empty.

> The Pentagon has now authorized a superior DNA-identification system, for which it will collect blood and saliva samples from all service personnel. (*Time*, January 13, 1992)

The long-awaited NAS report, *DNA Technology in Forensic Science*, finally arrived in early April 1992. It was seven months overdue, having been plagued by a threatened minority opinion, accusations of possible conflicts of interest, resignations, and leaks of confidential drafts. A *New York Times* front-page headline announced on April 14, two days before the report's scheduled release: "Judges Are Asked to Bar Genetic 'Fingerprinting' Until Basis in Science is Stronger."

At a hastily arranged news conference the same day, Victor McKusick, a world-renowned geneticist at Johns Hopkins University and chairperson of the report's panel, vehemently denied that the report had said anything of the kind. The next day, the *Times* front page carried a retraction of the previous day's headline.

When one reads this 185-page document, published by the National Academy Press, one finds a Summary Statement endorsed by the entire committee. It says, in part:

> We recommend that the use of DNA analysis for forensic purposes, including the resolution of both criminal and civil cases, be continued

while improvements and changes suggested in this report are being made.

That recommendation seems crystal clear. The next paragraph goes on to state, in part:

> We regard the accreditation and proficiency testing of DNA typing laboratories as essential to the scientific accuracy, reliability, and acceptability of DNA typing evidence in the future. . . . After sufficient time for implementation of quality-assurance programs has passed, courts should view quality control as necessary for general acceptance.

Critics were quick to point out that since no existing DNA testing laboratory could meet the standards of quality control and accreditation, it was difficult to see how the report could endorse the continued use of an admittedly flawed method of presenting evidence. But committee member Eric Lander said that, given the diverse nature of the 12-member NAS committee on which he served, "I think that this report does define general acceptance within the scientific community."

Certainly one notable caution offered by the report relates to the accuracy of estimated probabilities. They strongly urged the adoption of a "ceiling principle." This means that labs must establish the upper limit, or "ceiling," of the frequency of each allele at each chromosomal site analyzed in 15 to 20 genetically homogeneous populations, such as English or German. They would do so by collecting blood samples and establishing permanent cell cultures from 100 individuals in each population. When calculating the odds of a match, the lab should use the highest frequency of the allele found in any of the populations, or 5 percent, whichever is higher. This would tone down the estimates closer to 1 in several hundred thousand or a million.

The committee called for the establishment of an *ad hoc* expert group, a National Committee on Forensic DNA Typing. The committee would develop new approaches for judging the admissibility of DNA evidence without interminable pretrial hearings, as well as oversee the population blood studies and advise the courts on statistics. This proposed committee would be housed in the National Institutes of Health.

The report also cautioned against establishing a national DNA profile data bank on felons convicted of particular violent crimes until pilot studies confirm the value of such a system. The writers urged the formulation of strict legal guidelines for allowing access to such data, including criminal penalties for abuses. For reasons of both privacy and confidentiality, they considered the maintenance of DNA data on members of the general population (as is now done with fingerprints) inappropriate.

Their caution reflected a continuing concern about the potential for breaching the confidentiality of individuals' DNA records. We mentioned earlier that the military intends to use DNA profiles as the ultimate identification for its personnel. The Pentagon intends to collect DNA samples from "blood and oral swabs" taken from all active service members by the year 2000. The armed forces insist that the samples will not be tested for AIDS or drug use or used for anything else but identification. However, the military would have to release these records in the event of a subpoena issued as part of a criminal investigation.

Some have proposed that all newborns be DNA typed, which could later aid in the identification of a kidnapped child. A national database would slowly emerge. These data, linked with enough personal information would be a highly tempting database for use, for example, by scientists understandably eager to look for correlations between DNA profiles and diseases.

Nachama L. Wilker is the executive director of the Council for Responsible Genetics, a Cambridge, Massachusetts, based national organization working to see that biotechnology is developed safely and in the public interest. Wilker calls the military's plans to establish a repository of genetic information "an invasion of privacy on a level we have not before experienced." She likewise is concerned that leaks of DNA data could result in discrimination against former inmates. She urges that states should be allowed to gather DNA samples only from those convicted of serious crimes with a high repeat rate. So far, only five states have moved to bar any unauthorized access to DNA fingerprints. For example, in Wisconsin, companies are forbidden to use genetic data for employment or insurance purposes.

*   *   *

What does the future hold for the use of DNA typing in the courts? Despite the intent of the 1992 NAS report to end the controversy over the forensic use of DNA typing, the situation remains muddled. The states of California and Massachusetts and the U.S. District of Guam have ruled that DNA evidence is inadmissible using all but the most conservative statistics, despite the fact that the NAS report concluded that it should be accepted.

The objections center around the acknowledged scientific uncertainty over calculating the chances of DNA matches between individuals. The NAS report underlines the need for an expanded DNA database derived from subgroups of various populations. Until that is accomplished, the NAS recommends the use of the "ceiling principle" mentioned earlier, whereby very conservative judgments should be made on the chances of a DNA sample from one person matching someone else's DNA simply by random chance.

Even at odds of, let us say, 1 in 10,000 that two individuals would have closely matching DNA samples by chance alone, such evidence would seem at first glance to be very convincing that a defendant is guilty. But in the courts, defense attorneys often have succeeded in raising nagging doubts over the validity of these statistics. While scientists are comfortable with theoretical arguments, when a defendant's life hangs in the balance, an admission of uncertainty on the part of a scientist on the witness stand can be a very strong incentive for a judge to bar the DNA data as evidence, or for a jury to disregard it.

The consensus seems to be that this confusion is temporary. In the future, most agree, the introduction of DNA evidence in the courtroom will be routine. However, the proposed national committee on standardizing forensic DNA typing recommended in the NAS report has not yet materialized, and so far, Congress is not moving to establish such a body.

The fact that DNA typing is such a powerful tool to exclude the innocent is certainly a strong argument for its continued use. Current procedures will no doubt be placed by more sensitive and even automated technologies, perhaps rendering old concerns obsolete. As a matter of fact, Alec Jeffreys, that "DNA bloke" from Leicester, has only recently proposed a method of DNA typing that employs PCR copying of minute amounts of DNA, whose subsequent analysis can be automatically read out by computer as digital sequences.

Whatever the exact nature of the optimal system, it will have to be rapid, accurate, and inexpensive in order to be used on a large scale. Currently, each DNA profile takes about six weeks, at a cost of about $100 per sample. Actually, the ultimate identifier would be a way to determine the actual base sequence of segments of one's DNA. That can now be done, but the lack of sufficient data for comparison—and the expense— make it wholly impractical at the present time.

Meanwhile, by 1994 the FBI expects to have genetic profiles on 100,000 prisoners, most of them sex offenders, robbers, and other violent criminals. The FBI plans on testing a nationwide system linking all similar state DNA databases together, which will form a kind of genetic mug book. Police will search the book by computer, looking for suspects whose genes match the ones left behind at the scene of a crime in the form of blood stains, semen, hair, or saliva.

While an electronic search through this computerized catalogue of potentially damning DNA may one day be as routine as trying to match ordinary fingerprints is today, the related issues of privacy and confidentiality are certain to take on even greater urgency. The information lodged in a person's DNA reveals more than his or her personal identity. As the Human Genome Project unveils more and more specific DNA sequences linked to the presence of genetic diseases, or even predispositions to certain gene-based diseases, genetic databases will become even more tempting mother lodes of this personal information.

We face a future in which the very genes that design our bodies will be a matter of record. Society will have to decide who will have access to the unique, personal stories in our DNA.

Chapter 4

GENES AND DISEASE

Each of us arrives in this world quite by chance. The genetic baggage that we bring is not of our choosing. It has been packed for us by a complex of random, unpredictable events. One's father contributes a minute sperm cell, really nothing more than a nucleus propelled by a thrashing tail. The sperm respond to chemical cues and migrate through the mother's oviduct toward an awaiting egg. An egg is released into an oviduct once every few weeks. Each is only one from among the three million or so lying dormant in the mother's ovaries.

From the millions of sperm that surround the egg in a swirling stream, the first to pierce the delicate membrane of the egg cell triggers a rejection response, denying entrance to all others. The 23 chromosomes within the captured sperm nucleus emerge and mingle with the 23 already suspended in the large, nutrient-laden egg. This fertilized egg, or *zygote*, is now a chance assortment of forty-six chromosomes.

The cells which gave rise to the *gametes* (the sperm and the egg) were each the site of a random reshuffling of chromosomes. In certain cells of the ovaries or testes, during the complex process of forming gametes, the original 46 chromosomes first replicate to 92. These 92 are doled out into four gametes, so that each gamete receives 23.

The chromosome pairs separate, so that each gamete gets only one of each original pair (save for the occasional mistake). Which one it gets is unpredictable, so that a given sperm or egg cell might number among its twenty-three chromosomes anywhere from one to twenty-three chromosomes from each grandmother and grandfather. There are 2^{23}, or more than 8 million, potential chromosome combinations in the resulting gametes.

So we are each unique. Each of us begins as a zygote whose chromosomes bear a total of some 50,000 to 100,000 genes. These are the code, the instructions which within hours call for the first cell division. Then the two-celled embryo becomes four-celled, then eight, and so on as it grows in size and complexity. At each stage all of the enclosed genes in each of the cells somehow are signaled to become active or to remain dormant, awaiting their turn. By twelve weeks after fertilization, long since attached to the protective warm tissue which lines the uterus, the embryo has shaped itself into a miniature human organism. So immediately recognizable as human, though still minute, it is now called the *fetus*, meaning "the young one."

It is a hazardous journey from fertilization to fetus. Before we had highly sensitive methods to detect pregnancy in its very early stages, the frequency of miscarriage anytime after fertilization was considered to be about 15 percent. We now think that it is probably as high as 80 percent, including many embryos that, unable even to implant in the uterus, succumb silently and undetected.

Our genes are constrained to act out whatever their code dictates. Remember the central concept, which we have already introduced in Chapter 1: the gene code is the specific sequence of nitrogenous bases (A, T, G, or C) along one side or the other of a particular portion of the DNA double helix. Those sequences spell out instructions for the arrangement of amino acids in the proteins manufactured by the cell. Some of those proteins are structural elements in the cells, while others act as enzymes, driving the thousands of chemical reactions which enable life to continue.

Each of us then is different genetically from all humans who have ever lived or who will come in the future. And yet we all must share a great many genes with our fellow humans. After all, we need the same kind of genes to form and operate a human body. So we are united by a common humanness, which we each express in a unique way.

Our genetic baggage is packed by many rolls of the dice. Nonetheless, the selective forces of evolution have seen to it that most of us who have sufficient genetic belongings to survive after fertilization, implantation, and development in our mother's womb are prepared to confront the world into which we are born.

And yet we each arrive carrying, on the average, about a half a dozen dysfunctional genes. We may remain blissfully unaware of their existence until we or one of our close relatives are counted among the millions who suffer from a genetic disease.

*   *   *

At the turn of the century more people succumbed to tuberculosis than any other illness, and the next two leading killers were pneumonia and diarrhea. Since the widespread use of antibiotics after World War II such infectious diseases have given way to heart disease and cancer as the leading causes of death. We now have a wide array of antibacterial medications and of immunizations against a variety of bacterial and viral pathogens. Of course we need many more such weapons, as the race to combat AIDS so clearly demonstrates.

Genetic disease presents an entirely different scenario. The effect of, for example, sickle-cell anemia, muscular dystrophy, or Tay–Sachs, Huntington's, or polycystic kidney disease so far lend themselves only to palliative measures at best. Up to 30 percent of hospital pediatric admissions and 12 percent of adult hospital admissions are associated with genetic problems. These figures apply to the more privileged countries—in the poorer nations, malnutrition and its associated weaknesses are still the major scourge of children.

Almost 5,000 heritable disorders have been clinically characterized. Where have the causative dysfunctional genes come from? What are their effects? What can be done to diagnose their presence, fend off their effects, or prevent them from burdening our children?

Let's turn to the basics of genetic disease in order to understand just what such a condition means and what the implications are for diagnosis, treatment, and transmission to the next generation. Only recently have we been able to say very much about the exact nature of some of these diseases at their cellular and molecular level. Now, with the ever-expanding possibilities brought about by molecular biology, we are in the early stages of locating, isolating, copying, and analyzing the very genes responsible for some of these illnesses. We are taking direct aim at gaining some measure of control over our genetic inheritance.

GENETIC DISORDERS

Genetic disorders can conveniently be divided into four categories: chromosomal, single-gene, polygenic, and mitochondrial. Let's look first at some basic concepts underlying these disorders. This overview is by no means a comprehensive list of even the more common genetic diseases. We will emphasize instead the principles that flow from considering some prominent and intensively studied examples. We can then consider the implications of our currently available scientific knowledge on diagnosis and treatment as well as on the genetic counseling of parents, prospective parents, and family members.

Chromosomal Disorders

These include a wide variety of aberrations involving a decreased or increased number of all or part of one or more chromosomes. It also can mean a rearrangement of genetic material on one or more chromosomes. A chromosome is basically one long DNA molecule made up of genes and intervening sequences. Alterations in this architecture, depending on where they occur, may range from benign and unnoticeable to harmful or even lethal.

Most of these conditions cannot be predicted by analyzing the chromosomes of the parents. The chromosome abnormalities arise dur-

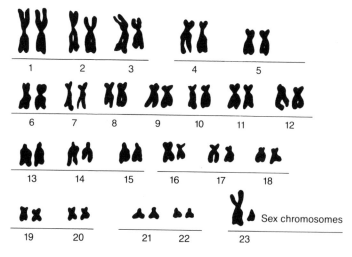

1 2 3 4 5
6 7 8 9 10 11 12
13 14 15 16 17 18
19 20 21 22 23 Sex chromosomes

Human karyotype

A *karyotype* of a human male. Note the larger X and smaller Y chromosomes. Each chromosome appears double because the karyotype is made after the chromosomes have doubled in preparation for cell division but before these newly doubled chromosomes have separated.

ing the formation of the sperm or egg or at some early stage of development.

Only one of the twenty-three pairs of chromosomes in human cells decides the sex of the individual. These are the so-called sex chromosomes, XX in the female and XY in the male. All of the 22 other chromosome pairs are referred to as *autosomes*. These designations are based on the *karyotype*, or actual appearance of the chromosomes under the microscope (see illustration). After the chromosomes in a cluster of dividing cells have been arrested in the act of replicating by chemical treatment, they can be spread out on a microscope slide, stained, and viewed. Each chromosome pair has a unique size and shape and so is easily examined and compared.

The gain or loss of one or more whole chromosomes is easily detected by a karyotype. Loss of a chromosome generally results in death to the fertilized egg. A notable exception is with Turner syndrome, which may occur once in about 3,000 female births. The affected individual usually has only one sex chromosome, an X. She is generally infertile, is short, and has kidney abnormalities.

The most common chromosomal abnormality among children is trisomy 21, or Down syndrome, in which the child has an extra autosome added to the 21st pair. This has an incidence of about 1 in 700 births, depending on the age of the mother. People with Down syndrome have various degrees of mental retardation, characteristic flattened features (the basis of the now-thankfully-obsolete designation "Mongoloid"), and, often, heart defects. Recent studies have shown that the extra chromosome can be traced to the mother's egg in about 95 percent of the cases and to the father's sperm in the other 5 percent.

One of several possible variants of Down syndrome is the situation in which the child has a karyotype of 46 chromosomes. Closer inspection reveals that one of the #14 chromosome pair has an extra #21 chromosome attached to it. This is a chromosome *translocation*.

In about half the cases, the parents have normal karyotypes, indicating that the translocation occurred during pregnancy. In the rest, one parent, though showing no symptoms, has 45 chromosomes, one of them a 14/21 translocation chromosome, which has been passed on to the child.

In Klinefelter's syndrome there is an extra X chromosome in each cell. The individual is XXY, rather than the normal XY. People with this syndrome are male (which shows that maleness is determined by the Y chromosome) and are infertile, with various mental and behavioral abnormalities.

In some instances, pieces of chromosomes may be missing. For example, some infants are born with the rare *cri du chat* syndrome, in which the deletion of a small segment of chromosome #5 is expressed as severe retardation, abnormal development of the head and face, and a shrill, plaintive, cat-like cry. The peculiar cry lasts for only a few weeks, but a significant number of affected individuals survive to adulthood.

Other deletions of chromosome pieces may be associated with malignancies. A missing piece of chromosome #13 is linked to retinoblastoma, a tumor of the eye which occurs in young children. A particular deletion on chromosome #11 causes Wilm's tumor, a childhood cancer of the kidney.

Chromosomal aberrations are usually lethal to the embryo and thus are never detected. Research using molecular techniques for those conditions described above and many others is beginning to reveal clues to their genetic basis. Each unique chromosomal defect has a particular pattern of physical symptoms. A specific region on chromosome #21 in individuals with Down syndrome has been narrowed to a segment containing from 50 to 100 genes.

However, because chromosomal disorders affect large segments of DNA, and therefore many genes whose specific identity and function are presently unknown, such genetic diseases are not now likely candidates for treatment or cure.

Single-Gene Disorders

More than 3,000 human diseases are known to be caused by a defect in one gene. Up to 8 percent of hospital pediatric admissions are due to single-gene disorders. Most are exceedingly rare, while others, such as cystic fibrosis, are much more common. In all cases, the symptoms are caused by too much or not enough of a protein or polypeptide (a protein building block) made according to the instructions in the code contained in only one gene. This results in defects due to missing proteins in vital cell structures or to missing or abnormal enzymes needed in metabolic pathways.

We have already seen that chromosomes occur in pairs. These pairs are not attached in the cell, and actually come in contact with each other only when the cell happens to be undergoing subdivision into sperm or egg cells. It is essential to understand that the members of a chromosome pair are *not* identical; they are similar in size and shape (that is, *homologous*), allowing us to diagnose gross, visible abnormalities, but at the submicroscopic level they are revealed to be quite different.

Scattered along the length of the DNA molecule that makes up each chromosome are the genes, specific sequences of DNA which code for proteins. Most of the DNA in our chromosomes is not genes at all, but a quite mysterious series of long sequences whose function is presently unknown. The individual genes are interspersed among these enigmatic stretches of DNA like so many black beads (the genes) arranged randomly in a long chain of predominantly white beads (the rest of the DNA).

On members of a homologous chromosome pair, genes at the same site, or *locus*, on either chromosome code for proteins that affect the same biochemical and developmental process, also referred to as a trait. These genes may take different forms, called alleles. Sometimes, the alleles for a given trait on the same locus of homologous chromosomes are identical, in which case the alleles are said to be *homozygous*. If the alleles are different, they are *heterozygous*.

If we continue our analogy of chromosomes as chains of beads, two homologous chromosomes lying next to each other would show a matching pattern of black beads (the genes, or alleles) among the more numerous, noncoding white beads. Wherever two black beads lay opposite each other, each would be a different shade if heterozygous, and the identical shade if homozygous.

Moreover, some genes are *dominant* over their corresponding alleles. The other allele would be considered as *recessive*. In that case, the dominant allele would be expressed, despite the presence of the recessive allele. The latter would only be expressed (i.e., make a protein) if it were in the presence of another recessive. In other words, dominant genes, whether in a homozygous or heterozygous pairing, are expressed. Recessive genes must be in the homozygous condition in order to function.

Please note that these are generalities whose exceptions are beyond the scope of this book. Given these few general principles, however, we can understand most instances of the inheritance of single-gene disorders.

Autosomal Dominant Disorders. Huntington's disease, which is transmitted in 1 in 2,500 births, is a particularly tragic example of such a condition. The symptoms usually do not appear until adult life, usually after the age of 30. By that time children have often been born before the devastating signs begin in the parent. Intellectual functions slowly deteriorate, while involuntary movements of the muscles eventually make control of bodily motion impossible—including swallowing. It leads inexorably to the person's death, often after many years of severe disability.

In another example, polycystic kidney disease, the adult form usually does not manifest itself until after the age of twenty. Then, cysts in the kidneys gradually increase in size, eventually ending in complete renal failure ten or twenty years later.

Other prominent examples are familial hypercholesterolemia, in which there is an abnormal increase in blood cholesterol leading to early coronary artery disease, and myotonic dystrophy, a weakening of the muscles and other abnormalities which can occur at any age and is extremely variable in severity.

All of these conditions are typical of autosomal dominant disorders in that symptoms do not appear until years after birth and the severity of the symptoms can vary widely from one person to another. But we no longer need always wait for symptoms. In Huntington's and polycystic kidney disease the presence of the dominant gene can now be detected, even prenatally. The ramifications of such a possibility will be discussed below.

We are looking here at a situation in which one abnormal gene dominates the other allele, its normal counterpart (see illustration). Remember that in sperm or egg formation one or the other of each homologous chromosome pair ends up in each gamete. Therefore, each child of a parent with an autosomal dominant disorder has a 50 percent chance of inheriting the dominant gene, and with it the disease.

Autosomal Recessive Disorders. Some diseases require the presence of a pair of abnormal recessive alleles. The parents of an affected child are usually carriers, that is, each has one defective gene but is protected by the presence of a normal allele (see illustration). Each child of two carriers has a 50 percent chance of being a carrier like the parents and a 25 percent chance of inheriting both recessive genes and thus the disease. These conditions affect males and females in equal proportions.

Familiar examples include sickle cell anemia, cystic fibrosis, Tay–Sachs disease, and phenylketonuria, or PKU. Some of these disorders are

Dominant Inheritance

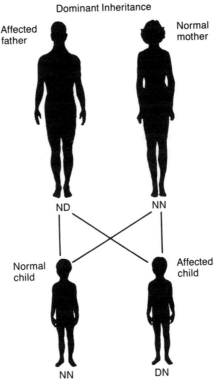

In an *autosomal dominant* disorder, each child has a 50 percent chance of inheriting the dominant disease-causing gene (D). Our example shows the dominant gene coming from the father. It could also come from the mother if she happened to have a dominant gene.

among the most common of all inherited diseases. In sickle cell anemia, a defective enzyme produces abnormal hemoglobin, the oxygen-carrying pigment which fills the red blood cells (see Chapter 1). Under any stress the cells assume a distorted, rigid shape. Few people with this disease survive beyond age 40. Death typically results from repeated infections, blood clots, or kidney failure. The disease occurs in 1 of 400 African-

Recessive Inheritance

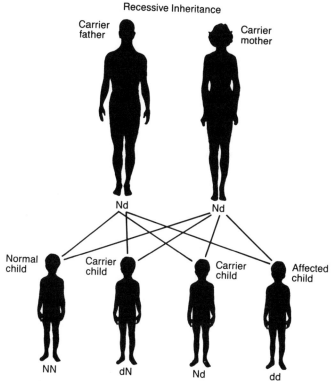

Carrier father

Carrier mother

Nd   Nd

Normal child

Carrier child

Carrier child

Affected child

NN   dN   Nd   dd

In *autosomal recessive* disorders, each parent is usually a *carrier* of one recessive gene. Each child has a 50 percent chance of being a carrier and a 25 percent chance of inheriting both recessive genes and thus the disease.

Americans. Carriers are easily detected. The carrier frequency can be as high as 1 in 8.

Cystic fibrosis, characterized by thick lung secretions, pulmonary infections, and impaired function of the pancreas, is nearly always fatal by the end of the third decade of life. Despite its lethality, approximately 1 in every 22 white Americans (and 1 in 17,000 African-Americans) is a carrier, and within the white population this disease is seen once in every

1800 births. Within a certain margin of error, the defective gene can be detected in carriers as well as in those with the disease—again, even prenatally.

Tay–Sachs is a lethal inherited disease. Children born with this disorder appear healthy at birth, but before the age of one they begin to show signs of central nervous system degeneration. Progressive mental retardation, blindness, and loss of muscular control follow, and the child dies at three to four years of age. The gene and carriers can now be detected. This condition is most common by far in families of Eastern European Jewish origin—the carrier frequency within this group is 1 in 30, and the disease strikes 1 child in every 3600 births.

In PKU the recessive alleles cannot code for an enzyme needed to break down the amino acid phenylalanine. Abnormally high levels of this acid build up, usually resulting in mental retardation. The fundamental flaw in this disease has been known for many years, and it can be routinely detected by simple biochemical tests shortly after birth. If diagnosed, a special diet low in phenylalanine will ward off the harmful effects. The incidence is 1 in 16,000 births, but is much rarer in Eastern European Jewish people and African-Americans.

Recessive disorders are generally much less variable in the severity of their symptoms than are the dominant disorders. However, in some cases of cystic fibrosis the person may be spared lung or pancreas problems through early adulthood, and some individuals with PKU may have normal intelligence but all of the biochemical aberrations.

On the subject of variability, there have been recent surprising findings which have added to the complexity of genetic disease inheritance. In 1988 Arthur Beaudet, a geneticist at Baylor College of Medicine in Houston, found a case of cystic fibrosis which had been caused not by the biparental contribution of one recessive gene apiece, but because both #7 chromosomes had been inherited from the mother. This phenomenon has subsequently been seen in other genetic diseases such as hemophilia (see below). Research may reveal that this scenario is much more common than we had ever imagined, and will complicate predictions of the odds of inheriting these diseases.

Also, we are just beginning to realize that in many cases, the expression of genetic disease traits depends on whether or not the disease genes are inherited from the father or the mother. This concept, which at first seemed almost heretical, implies that the very same gene may act differently depending on its paternal or maternal origin. This phenomenon, termed *imprinting*, is apparently controlled by genes which in turn modify the actual disease genes, most likely during sperm and egg formation. Investigations into imprinting is a major challenge for future genetics research.

X-Linked Disorders. These affect males and females differently because the genes responsible are on the X chromosome. Typically in X-linked disorders the mother has a defective gene on one of her X chromosomes (see illustration). Each of her children has a 50 percent chance of inheriting that gene, but the female child will usually be a carrier while the male child will develop the disease. The normal allele on the daughter's other X chromosome produces enough normal protein to make up for the defect. She is protected just as the mother is. The male Y chromosome, however, is not completely homologous with the X. Most of the genes on the X do not have corresponding alleles on the Y and can be expressed even if they are recessive, as is the case in most of the X-linked disease situations. X-linked traits cannot be passed on from father to son, because a male contributes only a Y chromosome to his male offspring.

Duchenne muscular dystrophy (DMD) and hemophilia are among the better-known X-linked disorders. DMD, the second most common of the nine forms of muscular dystrophy, has an incidence of 1 in 3,500 males. The presence of the defective gene results in minimal or absent levels of the muscle protein dystrophin. Most people with this condition succumb to cardiorespiratory problems or infections by 20 years of age. It begins as muscle weakness in boys three to seven years old, and within a few years they are no longer able to walk.

The specific gene responsible has been traced to its location on the X chromosome and its base sequence has been analyzed. It is the largest known human gene, covering some two million bases and containing some 60 subunits. DMD can be diagnosed before birth, and with some exceptions female carriers can be identified.

Hemophilia occurs because of a lack of Factor VIII (in hemophilia A) or Factor IX (in hemophilia B). These factors are proteins vital to two

X-Linked Inheritance

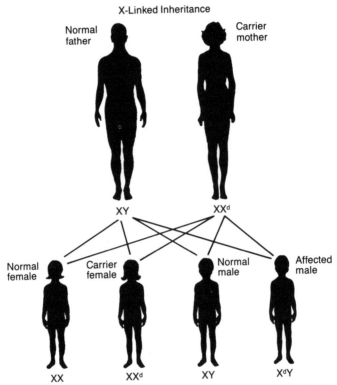

In *X-linked* disorders, the gene responsible for the problem is on an X chromosome. A female child, who inherits two X chromosomes, is protected from the effects of the gene. A male who inherits the disease-gene-bearing X chromosome along with a Y develops the disease. Males cannot be carriers in the X-linked situation.

of the many steps leading to normal coagulation of blood after injury. The incidence is 1 in 10,000 males for type A and 1 in 70,000 for type B. In either case, prolonged bleeding may occur, either spontaneously or provoked by even a minor irritation. The large genes for both hemophilia types have recently been found within the X chromosome and isolated.

By far the most common cause of mental retardation next to Down syndrome is a condition with a puzzling pattern of inheritance. Fragile-X syndrome occurs in about 1 in 1,000 males and 1 in 2,500 females. Its severity ranges from mild learning disabilities to such severe retardation that the child can barely talk or function.

This condition, first described in 1969, earned its name from the appearance of the X chromosome, in which one end of the chromosome seems to be hanging on by a thread. Twenty percent of the males with this "fragile" site are normal and are called "transmitting males." All the daughters of such a male will inherit the fragile chromosome but will be normal. However, their children, male or female, may be affected. In 1991, after years of extensive research, the gene which causes fragile-X syndrome was finally isolated.

Polygenic Disorders

Unfortunately for ease of genetic analysis and possible treatment, some of the most common genetic disorders have complex origins. Many of the chronic diseases of adults, such as diabetes mellitus, hypertension, and coronary artery disease arise from defects in an unknown number of genes. These disorders also include many birth defects like congenital heart disease, cleft palate, cleft lip, and spina bifida. Polygenic disorders arise from defects in the sperm, egg, or both, but at present the number and chromosomal location of the defective genes are not known.

Mitochondrial Disorders

This category of genetic disease is so new that it is not included in most recent textbooks. We have known for over twenty-five years that DNA can also be found in the cells' mitochondria as well as in the chromosomes. The mitochondria are often referred to as "powerhouses" or "energy factories" for the cells. They are minute, sausage-shaped objects in the cell's cytoplasm whose vital job it is to take energy from our food and convert it to the high-energy molecule ATP, which is used to power the rest of the cell.

The entire DNA sequence of the mitochondria has been studied intensively. It consists of almost 17,000 base pairs on a continuous loop of DNA. This is in marked contrast to the enormous chromosomes, whose DNA spans some 3 billion base pairs. Mitochondrial DNA codes for only thirteen of the proteins needed for cellular energy gathering. The other fifty or so are coded from the blueprint in the chromosomal DNA. Small though it may be, the mitochondrial genome can play a role in genetic disease, a role which we are just now beginning to explore.

Cells that need the most energy—such as those in the brain, heart, or skeletal muscle—have the greatest number of mitochondria. Some muscle cells, for example, may have as many as 10,000 mitochondria, as opposed to the few hundred found in skin cells.

Because the mother's mitochondria pass directly from one generation to another, and the mutation rate in mitochondria is much higher than in chromosomes, mitochondrial DNA has proven to be an excellent though problematic means of tracing the evolution of humans and other species (see Chapter 2). Apparently some of these mitochondrial mutant genes can cause disease, however. In 1989 Douglas Wallace and his colleagues at Emory University in Atlanta, Georgia, identified Leber's optic neuropathy as the first disease traceable to mitochondrial DNA. This very rare condition is marked by a gradual loss of vision as cells in the optic nerves gradually die. In all cases a single mutation in mitochondrial DNA was found to be the common denominator. This was a stunning discovery which defied the conventional wisdom that mitochondrial breakdown would end in death, given the critical nature of these structures.

Soon, however, as dozens of laboratories around the world began to concentrate on mitochondrial genetics, it became clear that this minuscule bit of maternal DNA was directly implicated not only in other rare genetic disorders but perhaps in more familiar ones as well. The details are difficult to sort out because some tissues have more energy needs than others, resulting in different patterns of disease in different tissues. One form of a disease may damage the brain and spinal cord, whereas in other people it may extend to other tissues such as the heart, kidney, or liver.

Scientists are hot on the trail of the intriguing possibility that mitochondrial mutations may be implicated in the degenerative conditions of Parkinson's and Alzheimer's disease. They may even prove to play a vital role in the aging process itself.

Mutations

Whatever the myriad effects of flawed genes, the assumption is that these gene permutations arise by one or more mutations, that is, heritable changes in the DNA molecular structure. Although the term may seem to bear a pejorative meaning, in living cells not all mutations are harmful. Some have no effect on protein structure at all, and other mutations may in fact be beneficial. After all, one of the basic mechanisms driving the process of evolution is the occurrence of mutations, which may confer an advantage to an organism's offspring.

At the complex, advanced level of human organisms, most mutations are neutral or harmful. We now know that mistakes are regularly made during DNA replication in cells and arise as well from environmental factors impinging on cells such as chemicals or radiation. In recent years scientists have discovered an amazingly efficient series of repair enzymes and other biochemicals which monitor DNA closely for defects and routinely correct them. In fact, one genetic disease, xeroderma pigmentosum, in which people are highly sensitive to light, arises from a mutation in a gene which codes for a protein used in the repair of DNA damage due to ultraviolet light.

Any effect an uncorrected mutation may have results from the fact that during the synthesis of proteins the base code in the gene is read linearly, always beginning at the same end. Mutations can disrupt this reading process, possibly altering the structure of the corresponding protein.

Gross mutations can occur whereby parts of the chromosomes are deleted, duplicated, or moved to another position. On a lesser but still damaging scale, the loss or addition of one or two extra segments to a gene can cause a so-called frame-shift, in which the whole code of bases following that mutation site are read out of phase. A totally different set of amino acids is built into the protein, destroying its function.

In a point mutation one base in the DNA is substituted for another. Since there may be different sets of three bases in DNA that code for the same amino acid, the results of a point mutation may be undetectable. On the other hand, in sickle cell anemia, the mere replacement of the code GAG for GTG results in abnormal hemoglobin in the red cells and the consequent devastating effects of this disease.

In tracing the pedigree of a genetic disease one needs to remember that a mutation causing that disease may be a "new" mutation. Some diseases may sometimes appear with no family history, such as the autosomal dominant condition of achondroplasia, a common form of dwarfism. It is now estimated that as many as 1 child in 20 can have a mutation not inherited from either parent that can be passed on to succeeding generations.

These mutant genes, which are passed on from one generation to the next, of course stay within the population that is interbreeding. That is why we see so many examples of genetic diseases such as Tay–Sachs or sickle-cell anemia confined almost exclusively to geographical areas or races in which genes are shared within a restricted pool. Humans have evolved as one species with the potential to interbreed. Geographical isolation, cultural habits, and other segregating influences have seen to it that certain populations sequester their pool of genes within their own ranks.

If these dysfunctional genes are often so harmful, how do they manage to survive and often spread throughout a population? One partial explanation is the carrier state, in which a gene may be silently borne from one generation to the other and detectable only through some elaborate diagnostic technique.

We now know that the presence of a sickle-cell gene confers a significant advantage to the carriers in regions where malaria is common. In South Africa, for example, the frequency of this gene is as high as 40 percent, apparently because the carriers' red blood cells are not a favorable milieu for supporting the growth of the malarial parasite. Those who have two of the recessive disease genes face an early death, but the life of those with only one (the carriers) is prolonged.

Some have theorized that the cystic fibrosis gene, very common in whites of European descent, may protect carriers from lethal bacteria-induced diarrhea, historically a leading killer of infants. Another intriguing theory suggests that the perils of insulin-dependent diabetes are partially offset by a genetic advantage that people with diabetes have which protects them during their fetal development against being lost by miscarriage. Perhaps genetic diseases are sometimes Faustian bargains with nature.

LOCATING DISEASE GENES

Genetic diseases, whether they arise *de novo* to strike without warning or manifest themselves as the inevitable outcome of a tragic family history, have been studied since the early days of this century. Their outward symptoms have been dutifully cataloged, and their often puzzling patterns of inheritance have been recorded. With the advent of DNA-based biology in the 1970s scientists were finally able to move from the clinical picture to the cellular, biochemical, and molecular details of these diseases that have diminished the lives of so many.

In so doing they have become caught up in a paradox. As the map showing the location of each of the 50,000 to 100,000 genes on human chromosomes fills in at a rapidly increasing pace, attention understandably has centered on those genes which cause disease (see illustration). Once pinpointed, the faulty genes can be isolated and copied by the millions in the laboratory. Their sequences can then be determined, and from them the proteins for which they code can be deduced. The reverse is true as well. The more classic approach of starting with the abnormal protein, when it can be found, can lead back to the gene sequence.

This gives us a hint at the biochemical pathways at the root of the disease's devastation. Moreover, having the gene in hand can be a powerful tool in probing for the presence of the disease gene in an embryo, fetus, child, or adult. Understanding the molecular details of the operations of disease genes may lead to methods to remove or diminish their influence. Uncovering their presence may allow an early opportunity to ameliorate their effects, give a basis for informative genetic

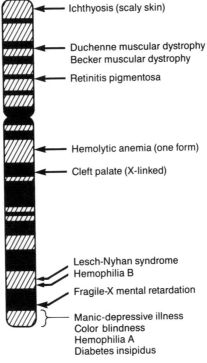

This map of the X chromosome shows the approximate position of the genes that cause the diseases listed. Similar maps are available for all the other chromosomes.

counseling to prospective parents, and provide the option of preventing the birth of children with serious genetic disorders.

The paradox lies in the widening gap between our increasing ability to detect the presence of these genes and our capacity to treat genetic disease, let alone cure it. However, we have no choice but to continue the search for the genes that bring about disease. By identifying the chromosomal location of the genes, and ultimately what proteins they make, in what quantities, at what time, and in what particular cells or tissues, we can hope ultimately to exercise significant control over them.

Before we point out the resulting ethical and societal dilemmas that are currently debated and will continue to be debated increasingly in private, professional, and public forums, it will be advantageous to understand the principles of the scientific detective work employed in tracking down hidden genes.

Metabolic Clues

We now know the defective protein produced in each of over 600 single-gene diseases. These include the well-known conditions of sickle cell anemia, PKU, Tay–Sachs, and hemophilia. We are still ignorant of the precise protein defect in the majority of genetic diseases.

Only a few genetic diseases can be treated because of our understanding of their chemistry. We can supply a missing protein such as Factor VIII in hemophilia, or design a diet for children with PKU in order to limit their intake of phenylalanine. For the most part, even when we know the biochemistry of the problem, there is nothing that we can do at present to alter its effects.

However, such knowledge sometimes can lead us to the gene itself. Lesch–Nyhan syndrome is an X-linked disorder marked by mental retardation and bizarre behavior, including compulsive self-mutilation. There is a missing enzyme (HPRT) in the cells of these children. Mouse cells which were known to produce excessive amounts of HPRT were grown in the laboratory. The abundant messenger RNA for HPRT was then isolated from those cells and used to synthesize the DNA of the gene, the blueprint for making that particular messenger RNA. The newly made DNA could then be used as a probe to pick out the same gene in human cells.

The use of molecular probes is a simple and extraordinarily important technique in DNA-based diagnosis, as well as in other facets of molecular biology. In Chapter 1 we described the molecular structure of DNA as a double helix. If one treats a sample of DNA so that the base pairs AT or GC separate, the two sides of the double helix are left as detached single strands. The bases are exposed like the teeth on a comb. Addition of any other single-stranded DNA with complementary bases (A for T, G for C, and *vice versa*) under carefully controlled conditions results in the joining of the added DNA (the probe) to the exposed bases. This *hybridization* of the bases results in a new double helix in the region where the probe has hybridized.

The probe zeroes in on that specific segment of the DNA which bears complementary bases, just as a magnet passed over a chain of beads will ignore the plastic ones but latches on to a metal bead when it reaches one.

In most cases the biochemical defects corresponding to a genetic disease are unknown. Other methods must be employed to locate the causative genes. A useful approach is to first seek out the locus of the gene on one of the 23 pairs of chromosomes. One way to do this is to fashion hybrid cells by fusing human cells with mouse or hamster cells. When the hybrids are grown in culture, there is a gradual loss of human chromosomes until one or only a few human chromosomes remain.

These can be broken into fragments by various treatments, and one is left with rodent–human cells which can be isolated and grown into a population of cells containing, in addition to the DNA normally found in the rodent, a small amount of human DNA. Probes prepared from known regions of human chromosomes can be applied to determine from what chromosomal segment the remaining DNA fragment originated. Whole panels of these human–rodent hybrids have been developed that represent overlapping segments of very small chromosome regions, allowing localization of genes to highly defined regions of specific chromosomes.

Chromosome Clues

Gene locations can be narrowed down by hybridizing probes to an individual's chromosomes spread out on a microscope slide. In order to visualize the hybridization, the probe may have attached to it a radioactive element or a fluorescent dye. After rinsing off excess radioactive probe, the slides can be exposed to a photographic emulsion which will show black dots where the "hot" probe has attached. Fluorescent probes are seen simply by viewing the slide under ultraviolet light. A fluorescent probe specific for chromosome #21, for example, can quickly reveal a Down syndrome trisomy #21 as three glowing spots in a scatter of chromosomes from one cell.

Sometimes the location of a disease gene can be discovered if the mutation happens to alter the appearance of the chromosome. The chromosomes can be stained so that they exhibit a characteristic pattern of numerous light and dark bands. This has been useful in a relatively few but significant instances, e.g., retinoblastoma and Wilm's tumor. For the most part, however, genes have been located by comparing the inheritance of a mutant gene with the inheritance of signposts, called *markers*, of known chromosomal location within a family.

Marker Maps

Mutations in DNA commonly result in nothing more than the substitution of one base for another. Since at least 95 percent of the length of a human chromosome does not function as genes, most of these changes are innocuous. Sometimes, however, these subtle changes alter sites in the DNA where restriction enzymes act. These enzymes are derived from bacteria. Each of the several hundred different restriction enzymes can recognize and cut through DNA wherever they recognize a specific base sequence, usually four to six bases in length.

This means that if one treats the DNA from a person's cells with one or more of these enzymes, the DNA is cut up into many fragments. The number and length of these fragments will almost always be the same for any one individual. A variation of this has become very useful as well as controversial in the identification of alleged criminals and for other forensic uses (see Chapter 3).

When a restriction enzyme site is altered in the DNA of one chromosome but remains the same in the other member of a homologous pair, the enzyme will not recognize the changed site. A longer fragment will be produced from that part of whichever chromosome has the altered site, because that particular piece will not be cut by the enzyme. This is a restriction fragment length polymorphism, or RFLP (see Chapter 3).

The individual is heterozygous for this RFLP, and because the RFLP is inherited just like a gene, we can follow it from generation to generation simply by recording the inheritance of the marker fragment. A central point here is that before the homologous chromosomes separate during sperm or egg formation, they wrap around each other and actually exchange segments. This is known as *crossing over* and results in *recombination*, that is, a joining together of formerly distinct chromosome pieces.

This phenomenon can be used to locate a particular disease gene. The closer the gene is to the RFLP site, the more frequently the gene and the site will be inherited together. The further apart they are on the chromosome, the better the odds that crossing over can occur, separating the gene from the fragment containing the RFLP site. If a particular RFLP almost always accompanies a disease in a family, the gene for this disease must be quite close to that fragment.

In 1980 several researchers championed the idea that RFLPs could be used to map all the genes in the human genome. The Human Genome Project to do just that has been under way since 1991.

RFLP analysis generally requires data from at least three generations of families with a genetic disease. In 1983 Jean Dausset and Daniel Cohen founded the Centre d'Étude du Polymorphisme Humaine (CEPH) in Paris. They make DNA samples from selected families available to interested researchers. The 40 families in CEPH have an average of more than eight children per family, and most have DNA from all four grandparents.

Central to the CEPH collection is the extensive contribution from families in Salt Lake City, headquarters of the Mormon Church. As an integral part of their religion, the Mormon leaders encourage their members to produce large families and also urge them to compile extensive genealogies—just what the doctor ordered for RFLP analysis. The CEPH families' DNA is immortalized as cell cultures, which can be grown and preserved indefinitely.

Once pieces of the genome near a gene of interest have been isolated, a variety of techniques can be used to determine the sequence of the DNA step by step until the actual gene is reached. Current methods for doing this are still tedious, so that more efficient means are urgently needed. As the map of the human genome becomes increasingly detailed through the efforts of the Human Genome Project, the lengths of DNA that need to be searched will decrease and allow us to pinpoint growing numbers of disease genes as well as all other genes.

There are some notable success stories even with our present limitations. Both RFLPs and the study of chromosome abnormalities led to the isolation of the gene for Duchenne muscular dystrophy, the X-linked disorder mentioned earlier that causes progressive muscle degeneration in young boys. The cystic fibrosis gene was found on chromosome 7 using RFLP analysis and sequencing of DNA from many families with a history of the disease. The breakthrough came in 1989 after four years of intensive searching by a large number of laboratories.

It should eventually be possible to track down any human gene. The task is an imposing one, as exemplified by the search for the Huntington's disease gene. Only in early 1993, after 10 years of intense research, was the defective gene finally located and identified. Equally challenging is the hunt for the genes contributing to many of the afflictions of humankind—e.g., coronary heart disease, Alzheimer's disease, schizophrenia, or manic–depressive (bipolar) illness. They arise from complex interactions among different genes and between genes and the person's environment, relationships for which we have few clues.

Assuming that we have captured partial or entire genes or have specific RFLPs in hand, what can be done to diagnose genetic disease?

GENES AND DIAGNOSIS

We can now diagnose more than 200 genetic disorders using DNA-based techniques. Many of these procedures are not yet routinely or widely used and have various margins of error, which must always be taken into account. They can be utilized when genetic disease is suspected—in a newborn with symptoms or one who has a parent with a dominant disorder or a mother carrying an X-linked disorder.

Of course, disease diagnosis can be made at any stage—prenatal, newborn, or beyond. These methods may also be used in a search for heterozygous carriers for recessive or X-linked genetic diseases or to identify individuals who have a dominant disease gene but are asymptomatic.

Considering the hardships imposed by so many genetic diseases, it is not surprising that much effort has gone into fashioning methods for prenatal diagnosis. When both parents are heterozygous carriers of a disease gene (a situation usually not discovered until after the birth of an affected child) the risk of having an embryo or fetus with the disease is 1:4. If the genome of a parent contains a deleterious dominant gene, for each child the chances that it is passed on are 1:2. If therapies are available, treatment can begin immediately after birth, such as dietary restrictions in the case of PKU. The parents are also afforded the option of abortion.

Prenatal detection of numerous genetic diseases has traditionally been based on tests to detect enzymes or other relevant proteins. More sophisticated biochemical analyses developed over the past few years have extended the range of this kind of prenatal testing. PKU, Tay–Sachs disease, sickle cell anemia and Lesch–Nyhan syndrome are all detectable by prenatal biochemical detection.

Prenatal testing of any kind requires either samples of amniotic fluid from the sac surrounding the developing fetus or direct sampling of fetal tissues. Some chemicals released by the fetus can be measured directly in the amniotic fluid, such as alpha-fetoprotein, which when present at elevated levels can point to spinal cord defects or at low concentrations may indicate Down syndrome. For the most part, cells are needed. A few shed fetal cells are found in the amniotic fluid. These can be collected and cultured for a few weeks to obtain enough biochemical products for analysis.

Amniotic fluid is withdrawn during amniocentesis by inserting a fine needle through the pregnant woman's abdominal wall and, guided by an ultrasound image, through the wall of the uterus and into the amniotic sac. Sampling cannot be done before the 14th or 15th week of pregnancy.

Fetal cells may be obtained earlier, between the eighth and twelfth week of development by chorionic villus sampling (CVS). The chorionic villi are fingerlike projections of the membrane surrounding the young fetus. They gradually develop into the mature placenta. Tiny samples can be removed through the cervix in sufficient quantities so that they sometimes may be examined without culturing. There has been a reported 2 to 4 percent chance of loss of the fetus following CVS as opposed to the 0.5 percent risk in amniocentesis.

A much less invasive technique for obtaining cells has been devised which may become the method of choice. In 1989, researchers in England and Italy reported the first molecular genetic prenatal analysis using maternal blood. They collected a few fetal cells from the mother's blood. The cells had managed to pass through the placental barrier separating the maternal and fetal circulations. The scientists amassed a sufficient amount of the cells' DNA by PCR amplification. In this case, they determined the sex of the fetus. Modifications of this technique may allow routine sampling for genetic defects as early as six weeks after fertilization.

It is now possible to perform even preimplantation diagnosis. It is feasible, though very expensive and technically demanding, to check on the genetic health of an embryo as early as the eight-cell stage. Allan Handyside and his colleagues at the Hammersmith Hospital in London reported in 1990 that they had determined the sex of such an embryo formed through *in vitro* fertilization. It has become a relatively common procedure to fertilize a woman's eggs with sperm in a laboratory dish and to implant the resulting embryos into her uterus (or much more rarely into the uterus of a woman who was not the egg donor). The London researchers succeeded in removing one cell from each of several eight-celled embryos, amplified their DNA with PCR, and used a DNA probe made from a Y chromosome to determine the sex of the embryos.

In clinical trials this procedure was used to detect male embryos of couples who were at risk of transmitting X-linked diseases, including fragile-X and Duchenne muscular dystrophy. Then only the female embryos were implanted.

Finally, in what one supposes has to be the ultimate step, Yuri Verlinsky of the Illinois Masonic Medical Center announced at the 1989 annual meeting of the American Society of Human Genetics that he had been able to search for a defective gene in an unfertilized egg, thus obviating the need to sample embryos and select only some of them for implantation. He had worked with a couple who were both carriers for a homozygous recessive disorder.

In the formation of a mature egg, there is an unequal cell division in which most of the cell contents remain in one cell, which becomes the egg, and a very small amount stays in a much smaller cell, which is discarded. Each has a full set of chromosomes. Removal of the small cell followed by PCR amplification and probe analysis shows whether or not the defective gene is in the smaller cell. If it is, then it can not also be in the egg. That is because in each cell of this carrier there is one recessive gene for the condition and one normal gene. If the egg is free of the recessive gene and consequently bears the normal gene, it can then be used for *in vitro* fertilization.

Prenatal diagnosis by looking for metabolic products has been limited because the product such as a flawed protein, may be expressed only in certain tissues. The product may not be found in amniotic fluid or cultured cells. We are also limited because we know of relatively few chemicals that are diagnostic of the many thousands of human inherited disorders. DNA-based techniques circumvent many of these problems and have brought about a revolution in genetic diagnosis. They focus directly on the structure of the DNA in the genome of the individual being examined. This means that only a few cells are needed as a source of that DNA, so that the kinds of DNA analysis made possible through the techniques of modern molecular biology are available for diagnosis literally at any point from fertilization to old age.

Prominent among those methods is the use of RFLPs. Just as these telltale DNA fragments may finally direct us to the chromosomal home of a long-sought-after gene, they can also simply detect the presence of that gene in a person's DNA.

The reliability of this diagnosis depends on the proximity of the locus of the RFLP cleaved by the restriction enzyme to the locus of the disease gene. The closer the two loci, the more reliable the diagnosis. For example, the Duchenne muscular dystrophy RFLP test is between 90 and 99 percent accurate. This can be improved in carrier detection by measuring levels of an enzyme that is elevated in 95 percent of DMD carriers.

In looking for Huntington's disease some laboratories have had to use up to ten variable loci detected by eight different probes and six different restriction enzymes. The accuracy varied between 95 and 99 percent. The most effective RFLP studies for Huntington's disease have needed a three-generation family that includes an affected grandparent and an affected parent. Because of this up to 40 percent of those who are at risk of this disorder have lacked sufficient family data. When the causative gene was identified in 1993, it became possible to test individuals directly. For some disorders common to certain ethnic groups (for example, the blood disease beta-thalassemia among Italians and Greeks), a set of RFLPs may be defined so that a diagnosis can also be made without analysis of family members.

Companies are already advertising services to families with a history of genetic disease. They will freeze cell samples, keeping them alive indefinitely and available for whenever genetic testing may require them. While grandpa may be deceased, his cells and their DNA can be stored for RFLP analysis and disease diagnosis of his descendants.

This complication introduced by multiple possible mutations in RFLP diagnosis applies to many genetic diseases in which there is a wide range in the number of alleles responsible for the same disease. Adult polycystic kidney disease has over 5 possible alleles, PKU has about 4, and cystic fibrosis has over 170. This raises serious questions about the efficiency and desirability of screening large populations for genetic disease carriers, because of the difficulty of determining if any of the many mutations are present.

Another useful way to detect mutations using restriction enzymes applies to the very few cases where the gene mutation directly alters the recognition site for that enzyme. The best example is sickle cell anemia, in which the mutation can be detected by digesting sickle cell and normal DNA and comparing the effects of a specific probe.

Likewise, in a very few cases a mutation may be spotted by using a probe that hybridizes directly to the normal or mutant allele. These allele-specific oligonucleotide probes (ASOs) are limited to the disorders where the base sequence of the alleles are known.

*   *   *

As the Human Genome Project opens up a cornucopia of genes for our study, the identification, analysis, and detection of disease genes will continue to proliferate. Future possibilities for treatment or even cure of genetic disease depends on the identification of the disease genes and a new understanding of their functions. Detailed chromosome mapping, now referred to commonly as *positional cloning*, is projected to lead to the identification of at least 100 genes responsible for human disease by the end of this century.

These future scientific advances will accelerate the current intensive debate over their implications. As the options for therapy lag even further behind the increasing capabilities for genetic analysis, serious questions arise. Diagnosis without available treatment leads to dilemmas.

Chapter 5

DNA, DISEASE,
AND DILEMMAS

DNA-based diagnosis may be a recent development, but as it has grown in sophistication and the spectrum of diseases which it touches broadens, there has been a concomitant increase in concerns about its medical, ethical, and social implications. The voices which have raised warnings about the perils that accompany the promises of genetic diagnosis have not been limited to academic journals. For example, the July 1990 *Consumer Reports* featured a five-page article, "The Telltale Gene," which pointed to "profound implications" for "health care—and your life."

Who could argue against the application of methods whose aim is to reduce human suffering? Is not genetic diagnosis like any other medical data, just another means of information gathering which allows one to make a medical or personal decision? There are more subtleties to the situation than one might suspect from a simple summary of the admittedly exciting advances in genetic science.

As a general background for our summary of some of the major issues surrounding genetic diagnosis, let's keep several points in mind. As new tests emerge, health care professionals often feel compelled to use them not only to increase their diagnostic capability but because the tests may afford more security against possible malpractice suits for failing to diagnose a given condition. Also, because DNA-based testing may be patented by biotechnology firms, there are strong financial incentives to establish those tests as a routine part of medical care.

Even given this dose of realism, can a responsible argument be made that since genetic diagnosis is ultimately meant to relieve human suffering it should be used whenever possible? In order to answer that question, we must make several important distinctions among the various situations in which genetic diagnosis can play a role.

DIAGNOSIS AND TREATMENT

Children may be born with a family history that indicates that they are at risk for a genetic disease or, lacking that, with symptoms pointing toward an inherited illness. Unquestionably, a diagnosis is needed so that the child may be given at least appropriate supportive care and so that the identity of the condition be known to the family to inform their future reproductive decisions.

Treatment of genetic disease can be general or supportive, such as that offered to any ill person as are antibiotics or physical therapy, or it can sometimes be directed at a particular disease. Unfortunately, the defective genes in question are present in all the person's cells and the effects often are widespread, poorly defined, and for the most part irreversible by current therapies.

For some disorders, avoidance of foods or other substances is helpful. People who are born lacking the gene for the enzyme glucose-6-phosphate dehydrogenase can limit damage to their red blood cells by avoiding many drugs such as sulfonamides. PKU, caused by another missing enzyme, is greatly ameliorated by a diet low in the amino acid phenylanine. Cholesterol-lowering agents may be used to treat the effects of the genes which escalate blood levels of cholesterol to dangerous concentrations. Kidney failure in the rare Lesch–Nyhan syndrome may be alleviated by the drug allopurinol. Sickle cell anemia requires frequent blood transfusions, while bleeding disorders may be controlled by administration of blood proteins. Bone marrow transplantation has been used to treat a number of disorders, including thalassemia and sickle cell anemia, while kidney transplants have helped in cases of polycystic kidney disease.

In this category of "diagnosis and treatment" we can also include polygenic diseases, those caused by complexes of genes, which may manifest symptoms later in life. For example, accurate diagnosis is needed so that diabetes can be controlled by diet and/or insulin injection. Identification of pituitary malfunction and replacement therapy with growth hormone can help children avoid pituitary dwarfism. This is a good example of how some gene products can be made in the laboratory and used as treatments for the negative effects of other genes.

In the search for what might be the ultimate weapon against defective genes, the development of therapies aimed at the genes themselves is already under way. For now, however, we usually are limited to trying to alleviate symptoms rather than offering a cure.

Preimplantation Diagnosis

Several *in vitro* fertilization (IVF) clinics both here and abroad already are offering preimplantation diagnosis. As described earlier, this procedure tests the DNA of a cell removed from an embryo. The embryo is easily available for manipulation because it is formed by addition of sperm to eggs in a laboratory dish. Embryos without the genetic defect searched for may be placed in the uterus for development. This approach is designed for couples who are at risk of having a child with a serious genetic disorder but who prefer not to wait until they may be faced with the option of abortion. Already, a healthy child has been born in England after this procedure showed that it was free of cystic fibrosis.

There have been several serious considerations raised concerning the use of this technology. First, one must understand that the use of IVF, while it has often allowed an otherwise infertile couple the opportunity to have a child, is an involved, expensive procedure, which has a relatively low rate of success. Eggs must be surgically removed from a woman after she has undergone stressful hormonal treatments, and the whole procedure more often than not must be repeated before there is a successful outcome. Only 14 percent of women deliver a live infant after one cycle of IVF—and each attempt costs an average of $5,000.

Because preimplantation diagnosis is still an experimental procedure, couples who agree to try this method will probably need to use CVS or amniocentesis as a backup to check on the health of the fetus, and they may therefore face a decision about abortion which they wanted to avoid in the first place.

In a lengthy special supplement devoted to the ethical concerns over genetic diagnosis in the July–August 1992 *Hastings Center Report*, Andrea Bonnicksen warned that as the number of diagnosable defects grows, embryo diagnosis could take on a momentum of its own, leading to the perception of this procedure as a means to a risk-free pregnancy. She also points out the perspective that despite the theoretical notion that embryo diagnosis represents the ultimate in disease prevention, it is yet another labor-intensive technology open to the few that can afford it "in an era of scarce medical resources and growing respect for community as well as in individual health concerns."

Prenatal Diagnosis

With little available treatment and no immediate hope for a cure for most genetic diseases, diagnosis of such disorders before birth offers limited options. Few issues are more controversial in American society today than abortion. Widely disparate views are argued with equal fervor. For the purposes of this chapter, we shall proceed on the assumption that abortion probably will continue to be an available legal option.

Prenatal diagnosis is considered preventive medicine. Principal indications for its use are considered to include a maternal age of 35 years or older or any of a number of family-history scenarios such as a known chromosomal abnormality in either parent or a previous child, an X-linked disease in a maternal brother or uncle, or parents who are both carriers of a recessive defect. Currently, almost all the prenatal genetic diagnoses done in the United States are carried out in the 5 percent of pregnant women who are over 35 years of age. Prenatal diagnosis has so far accounted for a reduction in the incidence of babies born with genetic disease of less than 5 percent. Many would advocate its wider application to prevent severe, untreatable genetic disease.

The primary reason that most women now have for seeking prenatal testing is to detect Down syndrome. As the number of genes discernible by prenatal testing increases the spectrum of detectable disorders, physicians will have access to much more information than ever before, including genetic information about many other traits of the fetus beyond its health.

What will limit the prenatal questions one will want to ask? Even well-established protocols are not ideal. Prenatal alpha-fetoprotein levels are derived from a maternal blood test which may indicate spina bifida, in which the spinal cord fails to close completely during development, and other neural defects. The test is becoming common despite the fact that it often does not give a definitive diagnosis. Positive tests need appropriate follow-up such as ultrasound or amniocentesis.

We should add as well that the severity of spina bifida, like that of other genetic diseases, cannot necessarily be predicted accurately by prenatal diagnosis. In Down syndrome, for example, one child may be healthy and only slightly retarded, while another may have a serious heart defect or be profoundly retarded. People with spina bifida can range from those with a very severe, life-limiting condition to people who live long, productive lives after surgical correction.

Prenatal diagnosis brings with it additional considerations. According to the August 27, 1992, *New England Journal of Medicine*, "every invasive prenatal procedure carries the risk of fetal death." They cite the risk of miscarriage after amniocentesis as 1 in 200. They also point out that chorionic villus sampling has led to subsequent limb and facial damage in newborns. A study of more than 3,000 pregnant women who have undergone CVS is now under way. Meanwhile, the authors maintain that "it is premature to place a moratorium on these new techniques until we can quantify the risks and weigh them against the benefit afforded by knowing whether the fetus is abnormal."

The current options after prenatal testing are few. One is to accept the possible outcome and prepare for it. Certainly there are families who seek prenatal diagnosis, not with the intention to abort a fetus with a genetic problem, but to prepare for their child's birth and postnatal care. With the diagnosis, the parents have the opportunity to seek out other families with a similarly affected child.

Another option is abortion. While abortion is a legal right, there are many for whom it is unacceptable on personal or religious grounds. Even for women who are not opposed to it in principle, the decision to abort can be a profoundly difficult one to make. Those who are opposed to abortion often regard those who choose to abort a fetus as insensitive people who regard life as having little value. While some of those on both sides of this issue may prove to be insincere or inconsistent, it may often be the case, as Barbara Katz Rothman points out in the same *Hastings Center Report*, that "the mother grieves over an abortion following prenatal diagnosis as she does the death of a child. For the mother it *is* the death of a child. . . ."

Rothman also points out that a woman who chooses not to have a prenatal diagnosis "has in some sense come to be seen [by society] as having 'chosen' to have a child with a disability. In choosing the risk, she is understood to have chosen the condition." There are several implications here. One is that as the number of prenatal tests increases, so does the sometimes subtle but always present pressure to abort. Of course, we are talking about a wide range of diseases, many of which may lead to an often brief life of suffering.

On the other hand, there are genetic diseases which, to be sure, are disabling but may diminish only certain of one's capacities during life. The term *disability* is central to a consideration of prenatal, or for that matter other forms of genetic diagnosis. Our society, quite simply, discriminates against people with disabilities. True, the Americans with Disabilities Act (ADA) of 1990 extends a comprehensive prohibition of "discrimination based on disability." This institutional prohibition was a hard-fought victory, but negative attitudes remain deeply entrenched.

Disability activist Mary Johnson maintains that "the desire to abort a 'defective' fetus says very little about the 'right to choose'; it speaks volumes, though, about our society's extreme reluctance to accept disability as a valid condition that does not signal a defective person, a defective life." Societal pressure to abort in some situations is increased because "disabled children face a hostile environment."

It should be noted as well that the problems of most children who are born ill or disabled but not with a genetic disease occur because they are born prematurely or poorly developed. For the most part this occurs because the mothers are living in poverty, are malnourished, smoke or use alcohol or other drugs, and often have no prenatal care. Much more money is spent on newborn intensive care and fetal diagnosis than on creating a social environment in which women could have healthier babies.

Genetic Screening

The possibilities for discrimination related to genetic disease extend beyond those related to disabilities. We should now make a distinction between genetic *testing* and genetic *screening*. The former is a general term which can refer to any protocol that looks for indications of the presence of a genetic disease, including the testing which we have already discussed. Screening, on the other hand, is the application of widespread genetic testing to large populations, prenatal to adult, including many individuals who are not known to be at risk.

Even assuming the advent of improved techniques to pinpoint even polygenic disorders like coronary artery, manic–depressive (bipolar), or Alzheimer's disease, would there be any advantage in large-scale detection programs? Are there any societal pressures which might lead to their implementation?

In the early 1970s the United States undertook screening for carriers of Tay–Sachs, which affects 1 in 3,600 Eastern European Jews, and sickle cell anemia, which strikes 1 in 400 African-Americans. The former screening program fulfilled its aims, while the latter was a disaster. The Tay–Sachs program began with a concerted effort to educate community and religious leaders. Testing was aimed principally at highly motivated young couples of childbearing age. Also, there was a prenatal test available to detect the disease. As a result, the number of newborns with Tay–Sachs decreased from about 75 in 1970 to 13 in 1980.

The larger-scale sickle-cell program lacked the resources and educational programs needed to enable people to understand the information they were given. However, some states passed laws requiring testing of newborns, school children, marriage license applicants, and even prison inmates. Many carriers mistakenly thought they had the disease itself, and some carriers were stigmatized and denied health insurance. Also, no prenatal test was available, so that some were pressured to choose abortion as the only way to prevent the disease, leading to charges of racism. The result was that few people availed themselves of the carrier test.

The Tay–Sachs experience, though limited to a small, distinct population, has been cited by some as a model for a mass screening for cystic fibrosis carriers. Cystic fibrosis is the most common disease among American whites. One in 22 is a carrier, and 1 in 1,800 suffers from the disease. However, the optimistic plans for carrier screening that accompanied the discovery of the gene responsible for CF was soon tempered by the growing number of different mutations which were found to be possible in the gene.

This DNA-based analysis thus became so complicated that in 1990 the National Institutes of Health recommended against a widespread screening program. The probes at that time identified only 70 to 75 percent of carriers and would detect only half the couples at risk of having a child with CF, leaving many uncertain as to their carrier status. The NIH recommended postponing consideration of any large-scale program until the test could detect at least 90 to 95 percent of carriers.

As of late 1992 the number of known mutations found in the CF gene had risen to over 170, although only about 20 of those are common. The accuracy of the testing had improved, making it possible to detect 85 to 90 percent of CF carriers. Pilot projects were under way in the United States, Canada, and Europe to study the appropriateness of the screening.

Another complication in CF screening is that as the ability to detect carriers has increased, so have the possibilities for therapy. In fact, now that we have isolated the gene and determined the gene product and its function, new and powerful methods of treatment for and possibly even control over the disease add a problematic dimension to the decision faced by two CF carriers. Should they choose to abort a fetus that has a disease that might be cured or controlled within a few years?

Technical efficiency is not the sole criterion for genetic screening. Who should be tested? Should CF screening be limited to whites, as the President's Ethics Commission recommended in 1983? At what age should testing commence? Should it be, as some have suggested, at the time when a couple gets a marriage license, missing all the unmarried couples who have children?

In 1992, an estimated 63,000 people in the United States underwent a test to determine their CF carrier status. Most did so because someone else in their family had cystic fibrosis. But some with no family history of the disease underwent testing because they hoped to have children, and CF is the most common genetic disease among Caucasians.

The most common screening technique is one which detects the 12 most common mutations causing the disease, but cannot identify the 15 percent of carriers with genes altered by other CF-causing mutations. Given the fact that some carriers still escape detection with the testing technology's current capabilities, the American Society for Human Genetics takes the position that the testing of persons with no family history of CF is "not recommended."

Screening programs as well as individual genetic testing requires careful, clear explanations of the results. A 1986 study of educated, middle-class, pregnant women revealed that 25 percent interpreted "1 out of 1,000" as 1 percent. Benjamin Wilfond and Norman Fost of the Department of Pediatrics and the program in medical ethics at the University of Wisconsin School of Medicine have calculated that in a scenario where screening would be done on three million people (about the number of yearly marriages in the U.S.) the required counseling time would be 651,000 hours. According to the National Society of Genetic Counselors, in 1992 there were about 1,200 counselors with master's degrees, with only 100 graduating to those ranks each year.

Precedents for screening programs do exist. The most widespread form of genetic diagnosis in the United States is newborn screening for metabolic disorders. Its goal is to identify children who might benefit from treatment before permanent damage can occur. It began in the early

1960s after Robert Guthrie devised a simple and inexpensive blood test for PKU. It is performed on a small blood sample collected on filter paper.

In the late 1960s and early 1970s Guthrie showed that the same blood sample could be used to screen for other disorders, including those in which genes for certain enzymes are defective. These deficiencies are not necessarily fatal, but cause mental retardation and other defects. Examples are galactosemia, homocystinuria, and maple syrup urine disease. The postnatal damage to the infant's brain can be at least partially prevented by modifications in diet. States vary in their mandatory newborn screening programs, but typically test for eight or nine of these genetic diseases.

With modern molecular analysis, a drop of blood is all one needs to extract and amplify the DNA within by PCR, opening up this more classic screening to many other possibilities. In 1992, C. Thomas Caskey, Director of the Institute for Molecular Genetics at the Baylor College of Medicine, forecast that at least a dozen genetic diseases would become amenable to DNA-based newborn screening. This possibility raises other serious concerns about genetic testing.

GENES AND DISCRIMINATION

Discrimination based on genetic information is the same in principle as any other differential treatment of individuals. It is based on real or perceived differences among them and is practiced for a wide range of reasons from prejudice to profit.

Distinctions may often be made justifiably. Habitual offenders may lose their driving licenses, or people with some dangerous infectious illnesses may be prohibited from food handling or contact with patients, both in the interests of public safety. Several of the major issues surrounding discrimination based on genetic conditions go beyond these straightforward issues. Two of the most contentious areas are those of employment and insurance.

Because of the current system of health insurance in this country, the two issues are inextricably linked. Before we treat these specific issues, let's first examine one newer aspect of DNA-based diagnosis, predictive testing, mentioned only briefly earlier in this chapter, which becomes highly relevant here.

Predictive Genetic Testing

Predictive genetic testing goes beyond the diagnosis of a symptomatic disease or establishment of carrier status. It is the detection of a gene or group of genes whose presence predicts, sometimes with certainty, and sometimes with quite uncertain odds, that one will at some time in the future develop a genetic disease.

Huntington's disease is an example of a serious, costly disease whose causative gene can be found in a person at any age. Other genes are being discovered and detected which predict not the certainty, but an increased probability, that one might develop a genetic disease such as a particular form of cancer, cardiovascular disease, diabetes, or any of several forms of mental illness. Some think that a tendency toward alcoholism can be detected by gene analysis, although that is highly controversial.

In the case of Huntington's disease it may be difficult for individuals to decide whether or not be tested. Does one want to know that years from now one will begin a prolonged, profound degeneration of the faculties? Does one want to test one's children for the gene? According to Dorene Markel, a genetic counselor at the University of Michigan Medical Center who deals exclusively with families with a history of Huntington's, very few people in which the disease gene is found are willing to conceive a child, knowing that each child would have a 50 percent chance of having the illness.

A recent discovery offers the opportunity for predictive testing in another rare disorder, which causes a family to be extremely susceptible to cancer, sometimes several types in the same person. This is called the Li–Fraumeni syndrome. Scientists have known for several years that a specific gene—the p53 gene—is critically important as a tumor suppressor. It appears to create a protein in cells that senses chromosome damage and helps to stop the cell that carries that altered chromosome from dividing out of control and causing a malignancy.

People with Li–Fraumeni syndrome inherit a mutant form of the p53 gene. The presence of this defective gene can now be detected. Carriers of the mutant gene have a 50–50 chance of developing serious cancer before the age of 30, and by 60–70 years of age their risk is 90 percent. Most people have two normal copies of the p53 gene. Unlike the situation with Huntington's disease, in this case testing of all family members offers some hope that early discovery of the harmful gene might save, or at least extend their lives. Early, regular mammograms, frequent rectal exams, and other careful monitoring could detect cancer at an early, treatable stage.

In the general population, diagnostic tests expected to be available in the near future for damaged p53 genes in cancerous growths could become widespread and important. Researchers are working on methods to detect such telltale genes in sputum samples, as early indicators of throat cancer, as well as for cells shed in stool samples, to detect early colon cancers. Others are looking at the possibility of screening women to see if their blood cells contain hereditary defects in p53, signaling a potential predisposition toward breast cancer.

The p53 gene is by no means the only candidate for predictive cancer testing. For example, another suspect is a gene found on chromosome 17, which may be responsible for breast cancer in some women before the age of menopause. But while predictive diagnosis may seem an ideal medical tool, it raises serious questions. For example, what will be the psychological pressures put on people who are found to carry a gene with possibly devastating consequences at some uncertain time in the future? Even if their test results remain confidential, would they feel compelled to reveal it if asked directly about their genetic history? And what about equal access to such expensive tests?

Within the next decade, we may find that clues now being uncovered about genetic predispositions to physical characteristics may lead to even more problematic issues about behavior. There are already controversial findings which are making newspaper headlines, hardly the occasion for dispassionate scientific analysis.

The usual occasions for scientists to test the waters with their latest theories and data to support them are scientific conferences. Such a conference, "Genetic Factors in Crime: Findings, Uses, and Implications," was to have taken place beginning October 9, 1992, at the University of Maryland. The meeting was cancelled after the National Institutes of Health withdrew its funding, charging that the conference too readily accepted the idea that crime and violence had genetic causes.

However, the conference prospectus noted that even if genes were one day found associated with certain tendencies even loosely linked to crime, such as impulsive behaviors, they would have "little specificity, sensitivity, or explanatory power." Dr. Gregory Carey, one of the researchers invited to the conference, emphasized in a September 15, 1992, interview with *The New York Times* that "There's no DNA segment that codes for crime. . . . My view is that the idea that there could be a genetic marker for crime is wild-eyed speculation or science fiction."

However, all biological activity, which includes behavior, ultimately does have a biological basis. That means, for example, that the genes governing the development of the brain and nervous system are prerequisites for their function. At this point the general consensus seems to be that one does not inherit "antisocial behavior." Perhaps certain personality traits are inherited. For example, Dr. Jerome Kagan of Harvard has found that there are genetic influences on certain types of temperament, such as shyness or a tendency to be uninhibited.

If future studies suggest closer links between genes and behavior than are now anticipated by most scientists, the possible social implications will bear close scrutiny. Society might be tempted to identify the bearers of any problematic "criminality genes" and seek to intervene to head off any anticipated criminal behavior. Most observers see merit in following the advice of the National Research Council's 1992 report on the future of violence research. They emphasized the need to study how social factors—poverty, neglect, or alcohol and drug abuse during pregnancy—affect the biological processes that cause the behavior.

Employment and Insurance Discrimination

Assuming that by genetic testing the eventual onset of one disease or another can be predicted accurately, and that the number of such diseases and the accuracy of testing will increase, what are the major concerns relative to employment and insurance?

Most Americans, about 150 million, have private group insurance, most often as an employee or an employee's spouse or dependent. Only about ten to fifteen million have purchased their own policies, while a total of 58 million are covered by Medicare or Medicaid. Thirty-seven million do not have health insurance.

Insurers, both individual and group, are faced with a common problem known as adverse selection. This means that, generally speaking, individuals aware that they are at a high risk are more likely to want substantial insurance. If insurers are not aware of these risks, particularly in the case of individual policies, they may have to pay a large number of claims, raising the costs of premiums, perhaps to a point where the system breaks down.

To offset this possibility, in plans which require detailed physical examinations, insurance companies may refuse or limit insurance to those with certain physical problems, such as those who test positive for the AIDS virus. In group policies the risks are pooled, and payments to individuals are shared among large numbers of premium payers. Nevertheless, premiums may be adjusted on the basis of the payout record of a particular plan.

Individual insurance covers only about 5 percent of the population, while employers bear the major burden of health care costs. Both insurance companies and employers have a vested interest in keeping health insurance costs to a minimum. As it becomes possible to analyze employees and prospective employees for the possibility of present and even future ills, decisions to hire and fire based on such information or to drop health insurance coverage altogether are logical steps, in strict business terms.

The Human Genome Project, the concerted international effort to map out all the hereditary information on the human chromosomes, began officially in 1991. The projections are for a 15-year, $3 billion effort to crack the genetic code of life. Along the way, genes are being discovered which play critical roles in disease. The organizers of this ambitious scheme are well aware of the ethical and social implications of this rapidly increasing pool of genetic information. According to James

Watson, we must legislate to make genetic discrimination illegal. "There are some things about which we must simply say you can't do it," he told geneticists at a 1990 conference in Leicester, England.

That is easier said than done. Accordingly, the Human Genome Project is devoting 3 percent of its annual budget to support the work of the Joint Working Group on the Ethical, Legal, and Social Issues Relative to Mapping and Sequencing the Human Genome. This is known, mercifully, as ELSI. In 1991 this committee, formed by the combined efforts of the National Center for Human Genome Research at NIH and the Department of Energy Human Genome Program, organized a Task Force on Genetics and Insurance. The Task Force was charged with studying "the likely impact of increased ability to predict future illnesses on insurance practices, especially access to health, life, and disability insurance."

One of the Task Force subcommittees, chaired by Paul Billings, is examining allegations that insurers are already engaging in genetic discrimination. Dr. Billings is a particularly appropriate choice. He is Chief of the Division of Genetic Medicine at Pacific Presbyterian Medical Center and a member of the Boston-based Council for Responsible Genetics (CRG). This is a national organization of scientists, physicians, public health advocates, and others who want to see genetics developed safely and in the public interest. One of its fundamental goals is to work to prevent discrimination based on information generated by the increasing uses of genetic analysis, including predictive testing.

In 1989, Billings and Harvard geneticist Jonathan Beckwith did a small-scale study of genetic discrimination in which they documented from 55 responders 29 who reported instances of discrimination by adoption agencies, employers, and insurers. These included three instances of health maintenance organizations that threatened to deny coverage because the subscribers were carrying fetuses with genetic defects. Abby Lipmann, an epidemiologist at McGill University in Montreal and a member of the CRG asks, "Are insurance companies going to define what is normal?"

In 1990 the CRG published a position paper on genetic discrimination which points out that "the use of predictive genetic diagnoses creates a new category of individuals who are not ill, but have reason to expect they may develop specific disease some time in the future: the healthy ill." They point out that there are few limits on what employers may use as preemployment medical tests.

The position paper states unequivocal support for laws that "prohibit discrimination in education, employment, insurance, housing, public accommodations, and other areas, based on present or predicted medical status or hereditary traits" and further propose "absolute and legally binding guarantees of confidentiality to protect information obtained from genetic screening."

On January 22, 1993, the Task Force on Genetics and Insurance wrote a letter to President Clinton, emphasizing its recommendations relative to genetic information and his intended health care reforms. The letter points out that as science continues to discover genes related both to existing disease and risk of future disease, there will be an increasing number of people who could be caught up in a dilemma. The genetic information in their medical records can "make it more difficult for both individuals and their relatives to acquire the health care financing and services they need."

The Task Force urges that "no one be denied access to health care financing or services on the basis of genetic information about their health risks." It likewise asks that people be allowed the right to give "free and informed consent" to any disclosure of their genetic information. Also, they recommend that health care programs should provide services for individuals and their families "with genetic disorders and those at genetic risk."

This leads us to a fundamental question. Is health care a fundamental right of our citizens? Is it the role of the government to see to it that each person has adequate health care, whether that means underwriting insurance programs or instituting a national, tax-financed health insurance program, perhaps similar to the Canadian system? That question will be argued with increasing fervor as insurance as well as health care costs continue to escalate, and the future offers us the means of locating any of hundreds of disease genes, now hidden in our genomes.

*   *   *

We have only just begun the inevitable progress in disease gene analysis and the private and public debate which will continue to accompany it. Networking and advocacy groups have been formed, such as the Alliance of Genetic Support Groups and The National Organization of Rare Disorders. Because of what we are learning about our genes, the worlds of science, medicine, and society are linked perhaps as never before. The accumulation of genetic knowledge grows faster than the understanding of how to use it wisely.

We are even beginning to develop the capability to alter our genetic endowment. We are on the threshold of a future in which we will be able to add genes to our genomes. Paradoxically, this problematic possibility has accompanied a mission of mercy, a victory in the struggle against disease genes—the medical miracle of gene therapy.

Chapter 6

GENE THERAPY

The frail young boy lay barely conscious on his bed in Houston's Texas Children's Hospital. David had been born twelve years earlier by cesarean section and immediately placed in sterile isolation. His parents had lost another son to SCID (severe combined immune deficiency), a rare genetic condition which renders the immune system virtually useless. David, too, had inherited SCID. Since he lacked the body's normal defense mechanisms, exposure to even a mild cold virus would be fatal. David was soon transferred to a sterile "bubble," the first of several plastic enclosures that would be his permanent homes.

David grew from a baby in a crib to an A student who attended school via two-way television. His clothing, food, books, and toys were sterilized before being passed to him through an air lock. He never felt the touch of a naked human hand or felt his mother's kiss. His only slim hope for a normal life was a bone-marrow transplant. Ideally this would come from someone whose bone marrow matched his so closely that his body would not reject it. His parents tried in vain for two years to find such a compatible donor. They decided to go ahead with an imperfect match from his healthy 15-year-old sister.

Several months later David developed diarrhea, vomiting, fever and dehydration, necessitating treatment outside the bubble. He crawled out of his sterile world into the waiting arms of his mother. For two weeks his condition worsened as his digestive system ulcerated and fluid accumulated in his lungs and around his heart. Finally David's young heart succumbed to the strain. It was February 1984.

In the Pediatric Intensive Care Unit on the 10th floor of the massive Clinical Center at the National Institutes of Health, a four-year-old girl sat patiently on her bed. She watched as a milky fluid dripped steadily through a tube inserted into a vein in her arm. She, too, was afflicted with a form of SCID, known as adenosine deaminase deficiency (ADA). Despite having been treated for two years with a new medication, she was now seriously ill. But a new age in medicine was beginning.

Suspended in the intravenous fluid were about a billion of her own white blood cells, which had been treated in the laboratory with genes. She had inherited a defective copy of a gene, and now normal copies had been inserted into some of her white cells. Over the succeeding months and years the genes would work to restore much of her immune system to normal function. Although she has had to return for periodic transfusions she can now face exposure to colds and infections in relative safety as she attends public school.

Her treatment was initiated on September 14, 1990, at 12:52 p.m. The era of gene therapy had begun.

*   *   *

As early as the 1960s the intriguing realization began to dawn that the exciting new developments in the technology for isolating and cloning genes might ultimately lead to their use as a means of combating disease. In 1967, Marshall Nirenberg, who was to receive a Nobel prize for his critical contribution to deciphering the language of the genetic code, predicted:

> My guess is that cells will be programmed with synthetic messages within 25 years. . . . Man may be able to program his own cells long

before he will be able to assess adequately the long-term consequences of such alterations. . . .

It had been a long and convoluted path from Nirenberg's uncannily prescient remarks and the first approved gene therapy clinical trial, described above, almost exactly 25 years later. The tale of the journey along that path is one of science and medicine combining forces in a highly technical, sometimes controversial effort to combat genetic disease.

Approximately 14 percent of newborn infants are afflicted with some form of inherited physical or mental problem. We have already seen that these include chromosomal abnormalities such as Down syndrome, over 3,000 single gene disorders like sickle cell anemia or cystic fibrosis, as well as problems caused by little-understood interactions between genes and the environment. The latter group includes the common conditions of spina bifida and juvenile onset diabetes.

With very few exceptions, particularly in the case of the monogenic (single-gene) disorders, modern medicine has been able to offer only treatment of the symptoms with no hope of a cure. More than half of all monogenic diseases lead to an early death. Those people who survive beyond infancy often face a lifelong struggle with the limitations imposed on them by their genetic inheritance. The genetic lottery which mingles genes from two parents into a new unique combination in their child has been beyond the reach of medicine. The genes have always spelled out our potential, which even under optimal conditions is inherently circumscribed by the quality of the messages in our DNA.

Are we always to be prisoners of our genes? Can we do more to alleviate this human suffering than the dietary, pharmacologic, and surgical interventions already discussed in Chapter 5, which offer some relief in varying degrees to some of those who are afflicted with genetic diseases? Beyond these, might we some day be able to prevent such genetic mishaps from occurring? Ultimately could we (and should we) develop enough control over genes so that we might remodel human beings into people who are free from all disease?

Consider the problem. Deep in almost all cells of the human body there is a control center, the nucleus. Hidden behind the membrane surrounding each nucleus are 46 chromosomes, long coiled strands of DNA spelling out the instructions in the 100,000 or so genes scattered along their twisted length. The actual functional genes make up only about 2 percent of the chromosome, while the rest is a puzzling patchwork of DNA sequences of as yet unassessed utility.

At any one time only a small fraction of those functional 2 percent is at work coding for the cell's proteins. The rest is turned off by a little-understood complex control system within the cells, a feedback system which sees to it that certain genes act only under specific circumstances. This means that in the retina of the eye, for example, a unique set of genes is operating which is dormant in the liver or the skin. This results in the variety of form and function necessary in the cells, tissues, and organs making up the complex human organism.

The microscopic genes are inaccessible to any kind of surgical intervention. Defective genes cannot be physically removed from the living cells and replaced by normal ones. The genes are there to stay. But could it ever be possible to ameliorate the effects of the defective genes not by symptomatic treatment of the entire body but by adding normal genes to the nuclei of at least some of the body's cells?

That is precisely what is now being done. Gene therapy is now in its early stages of clinical trials and development. In the space of a few years it has passed from being a distant ideal in the minds of a few pioneers to a reality. It is not, however, the only conceivable form of genetic manipulation in humans. There are four such possibilities:

1. *Somatic-cell gene therapy* aims at introducing genes into some of the somatic (body) cells to correct a genetic defect. This is the type of manipulation most commonly referred to as "gene therapy." In this case any effects would be limited to the person involved. The new genes would not be passed on to future generations.

2. *Germ-line gene therapy* would result in the correction of the genetic problem in an individual's reproductive cells so that it would no longer be passed on to his or her offspring. If performed early in the individual's life, perhaps even at the embry-

onic stage, the disorder might be corrected in the person being treated as well.

3. *Enhancement genetic engineering* would consist of inserting a gene at a pertinent stage in a person's development so that a specific characteristic would be improved or enhanced, such as the introduction of a gene to increase the amount of growth hormone production, resulting in a taller individual.

4. *Eugenic genetic engineering* might come about by inserting genes to change or improve complex traits which arise from the interaction of many genes with the environment, such as personality or intelligence. If either the eugenic or enhancement engineering were done at an early enough stage, such changed traits might be passed on to future generations.

By 1993 over 40 clinical experiments for gene insertion into humans, most in the United States, had been approved by an unprecedented series of intensive regulatory reviews. In addition to SCID, the diseases to be confronted include various forms of cancer, liver failure, hemophilia, high blood cholesterol, and even AIDS.

Enhancement genetic engineering and germ-line genetic manipulation are being carried out in animals. These experiments have given scientists the technical ability to apply some of these genetic modifications to human subjects. Eugenic genetic engineering, for now at least, is beyond our capabilities.

There of course exists a critical distinction between what can be done and what ought to be done. The four categories of genetic manipulation listed above open up a wide array of medical, legal, philosophical, and ethical concerns. What makes these questions unique is that they are raised against an often subliminal background of deep misgivings. Should we tamper with the very control centers of our cells? Do we dare, despite our best intentions, to alter our genetic inheritance or even to decide what that inheritance will include for the generations that will follow us?

A perspective on these vital issues can begin to form only in the context of a basic understanding of the history, science, and technology of gene manipulation. Only when these are understood can we adequately confront the debate on its promises and perils.

THE GENE DREAM

By the late 1960s it had become clear that certain tumor-causing viruses were capable of an extraordinary feat. They were able to integrate their genetic information randomly into the chromosomes of the cells they had infected. As these cells divided, the newly integrated genes were passed on along with the rest of the genome. The viral genes became functional and initiated the growth of a tumor. This immediately suggested the possibility of manipulating these viruses in the laboratory so that they could be used to introduce not their harmful genes, but other normal, useful genes, into otherwise defective cells.

There were several imposing technical obstacles, not the least of which was that no one yet knew how to isolate usable amounts of specific genes. Then, typical of the dizzying pace of advances in molecular biology by the early 1970s, the complete chemical synthesis of a yeast gene was reported as well as a method for efficiently transferring genes into viruses.

With these and other advances in mind, Theodore Friedmann and Richard Roblin from the Salk Institute in La Jolla, California, used the optimistic term "gene therapy" in a March 1972 *Science* article to refer to the proposed use of isolated DNA segments or viruses to treat humans suffering from genetic diseases. While championing further research directed at the development of techniques for gene therapy, they nevertheless stated emphatically that "for the foreseeable future . . . we oppose any further attempts at gene therapy in human patients. . . ."

They went on to cite the lack of understanding of the process of gene regulation in human cells as well as the rudimentary state of the understanding of the details of the relation between a genetic defect and the resultant disease. They called for a major effort to create a set of scientific and ethical criteria to guide the development and application of gene therapy techniques.

The call to halt further attempts at gene therapy referred to the fact that a first attempt had already been initiated. Two years earlier Dr. Stanfield Rogers, an American scientist working in Germany, had assisted in injecting three sisters with the reputedly harmless Shope papilloma virus. All of the young girls had suffered from the genetic disease argininemia, in which a single-gene defect lowers levels of the blood enzyme arginase. The ordinarily harmless amino acid arginine accumulates in the blood and spinal fluid, resulting in severe mental retardation.

Rogers knew that the Shope virus had been shown to restore adequate levels of arginase in tissue culture cells and theorized that perhaps it might do the same in humans. The fruitless attempts at treatments for argininemia which he carried out between 1970 and 1973 sparked an immediate debate that still resonates today. It extended to the halls of Congress in 1971 when Senator Walter Mondale called for a study of the legal, social, and ethical implications of geneticists' research as part of the mandate for a proposed National Commission on Health, Science, and Society.

In that same year a national conference, "The New Genetics and the Future of Man," was convened by Canon Michael Hamilton of the Washington Cathedral. There, Dr. W. French Anderson, who was to become the central figure in the development of human gene therapy over the next two decades, defended Rogers' actions as morally justified given the harmless nature of the virus and the serious, otherwise untreatable nature of the illness. Nevertheless he warned that "success here might encourage less justified attempts at premature gene therapy." He endorsed Mondale's proposal so that goals and safeguards could be established while public education kept pace with the anticipated scientific advances which would eventually lead to safe and effective therapeutic use of genes.

The optimism sparked by the dramatic advances in molecular biology in the late 1960s was justified as a series of new discoveries led to the means of dicing and splicing DNA almost at will. We have already given a brief account of these years in Chapter 1 where we noted the unprecedented self-imposed moratorium on recombinant DNA research called by scientists in 1973–1974. Concerns over the possibility of safety hazards led to subsequent regulation of DNA research by the National Institutes of Health Recombinant DNA Advisory Committee, which published extensive guidelines in 1976.

Buoyed by the rapid pace of progress in cloning and manipulating genes and with a greater sense of security that their concerns over safety were being met, scientists entered the 1980s with high hopes that the decade would usher in a bright future for effective gene therapy. The decade began, however, with a highly controversial decision by Dr. Martin Cline.

Dr. Cline, Chairman of Hematology at UCLA, had asked his Institutional Review Board to review his plan to treat patients with sickle cell disease and other blood disorders with genes linked to bits of bacterial DNA, which we described earlier as recombinant DNA or rDNA. He later revised the protocol to insert the genes alone. Prior to an IRB decision on the matter, Dr. Cline went to Israel and Italy, where he treated two patients with a severe form of thalassemia, a genetic disease marked by serious anemia. He removed some of their bone marrow, exposed it to rDNA containing normal hemoglobin genes, and transfused it back into the patients with apparently neither beneficial nor deleterious effects.

The procedures were carried out on July 10 and July 15, 1980. The UCLA Institutional Review Board disapproved the protocol on July 16. Dr. Cline explained that he had used the protocol "because I believed that they [the genes] would increase the possibility of introducing beta-globin genes that would be functionally effective. . . . I made this decision on medical grounds."

In the wake of disclosure of his actions, Dr. Cline resigned from his department chairmanship. NIH terminated several of his grants and stated that future grants would necessitate written compliance with NIH rDNA guidelines and human subject review processes. (Dr. Cline subsequently regained funding and went on to maintain a productive research laboratory at UCLA.)

With the notoriety of the Cline case, the notion of the practice of human gene therapy was elevated in the public consciousness from the realm of distant science fiction to the real world. Unfortunately for its serious proponents, this new perception had been born in scandal, hardly a venue for calm and enlightened discussions.

Ironically, just prior to Dr. Cline's foreign experiments, religious leaders of three large faiths—the National Council of Churches, the U.S. Catholic Conference, and the Synagogue Council of America—had written to then-President Carter regarding their growing concerns about rDNA technology and intervention with the human genome. Carter responded by asking his newly established Commission for the Study of Ethical Problems in Medicine and Biomedical and Behavioral Research to take up these issues. The Commission's response was the first major document by a governmental body to examine in depth the broad ethical implications of gene therapy. Their report, *Splicing Life*, was released in November 1982. It would set the agenda for the regulation and review of a breakthrough which would not take place for another seven years.

The Commission had begun its work in 1980, at a time when respected critics warned against "fundamental dangers" posed by altering human genes and the press carried analogies to Dr. Frankenstein's experiments, giving the impression that gene therapy would create monstrous human–animal hybrids. *Splicing Life* did not silence all of the critics, nor should it have, but it vigorously defended the continuation of rDNA research. It carefully distinguished between acceptable and unacceptable procedures in such research. Most significantly, the report saw somatic-cell gene therapy as not significantly different from any other standard medical intervention.

Moreover, it referred to such science as a "celebration of human creativity" while pointing out that if germ-line therapy were ever technically possible it would undergo intense public scrutiny before being undertaken. The authors made a series of recommendations for continuing oversight of human gene therapy. Some of these have since taken the form of:

1. a decision that the already established Recombinant DNA Advisory Committee (RAC) would take the responsibility for reviewing proposals for gene therapy on a case-by-case basis only after they had been approved by local Institutional Review Boards and Institutional Biosafety Committees;

2. the formation in 1984 of the Human Gene Therapy Subcommittee of RAC; and
3. creation, also in 1984, of a Biomedical Ethics Board composed of six Senators and six Representatives (the Board would end its activities in 1989 amid political infighting).

Meanwhile the cautious but strong support for somatic-cell gene therapy had quieted much of the concern about its possible dangers and misuse. Not so, however, for the attorney–social activist Jeremy Rifkin. The widely read and influential Rifkin, sometimes referred to by his critics as the "Abominable No Man," warned that "Once we decide to begin the process of human genetic engineering, there is really no logical place to stop." His advice was to forbid gene therapy, even at the cost of loss of treatment.

Rifkin's absolutist position was no longer held by mainline religious leaders, who by 1983 leaned towards supporting somatic-cell therapy. Switching his focus, Rifkin authored a resolution signed by 56 religious leaders (among them many Jewish, Catholic, and Protestant officials) and eight scientists, which was sent to the U.S. Senate. It stressed that "efforts to engineer specific genetic traits into the germ-line of the human species should not be attempted." In light of the renewed controversy, then-Representative (now Vice-President) Albert Gore requested that the congressional Office of Technology Assessment begin a study of human gene therapy in March 1984.

A panel of 17 scientists, clinicians, genetic counselors, family members, ethicists, historians, theologians, and public policy experts prepared an extensive report, which was reviewed by 70 outside experts before being released on December 10, 1984. Once again, somatic-cell gene therapy was judged not to be distinct from other medical treatments. They pointed out that it might in fact allow treatment of otherwise untreatable conditions. Apparently to avoid a protracted debate over the emotionally charged question of germ-line therapy the subject was sidestepped. After all, the state of the technology of the time offered little hope of even being able to efficiently insert a gene into an embryo or egg cell.

Meanwhile, behind all of the public discussion and debate, the science of gene manipulation was steadily moving ahead. Procedures were developed for inserting functional genes into the bone marrow of mice. Rat growth hormones were placed into marrow cells of dwarf mice, resulting in mice which grew to one and a half times their normal size (not exactly "giants," as reported in the press, but still nothing to sneeze at). *Retroviruses*, a unique form of virus readily manipulated in the laboratory, promised to be an effective gene-delivery system, based on animal studies. The stage was set for the first approved transfer of genes into human beings. It took place on May 22, 1989.

This groundbreaking procedure was technically not gene therapy. Genes that might cure a disease were not used. Instead, a bacterial gene which normally makes bacteria resistant to the antibiotic neomycin would be linked to certain of the patients' white blood cells as a marker to make it easier to track the fate of those cells. Steven A. Rosenberg and R. Michael Blaese, both of the National Cancer Institute, and W. French Anderson of the National Heart, Lung, and Blood Institute had submitted a proposed gene-transfer protocol on June 20, 1988, for approval by the Investigational Review Boards of the National Institutes of Health. Eleven months later, after having worked its way through seven regulatory bodies and fifteen reviews (including a final appraisal by the Food and Drug Administration), more than any clinical protocol in history up to that point, it received final approval. Five patients, all with advanced melanoma, a form of skin cancer, and life expectancies of less than 90 days, signed consent forms, and the test was under way.

Why the marker genes? The answer lies in the Rosenberg group's approach to cancer therapy. There are three traditional methods: surgery to remove cancerous tissue, radiation to shrink tumors, and chemotherapy, the use of drugs to destroy abnormal cells. Alone or in combination these cure about half of those with cancer. Yet at current rates one in six people in the U.S. and Europe will die of cancer, almost half a million people each year.

The body's immune system is not totally helpless against cancer. In fact, a certain class of white blood cells, the T lymphocytes, regularly detect and destroy cancerous cells before they can become established as a deadly growth. Also, cells of the immune system can control one another's activities by secreting powerful hormones known as cytokines.

One of these is interleukin-2 (IL-2), which stimulates T cells to divide. T cells are exquisitely specific; that is, they can recognize specific proteins (antigens) on a cell's surface and bind to them. They then multiply and kill the cell. Since cancer cells sometimes bear on their surfaces antigens not found on healthy cells, such cancer cells may be snared by T cells which recognize those antigens.

T cells can be grown in laboratory cultures with the added stimulus of IL-2. Rosenberg and colleagues reasoned that if they could isolate from a patient even a small number of T cells that would react against that person's tumor, large numbers could be grown and infused into the patient to attack the malignant growth. They postulated that if the immune system is in the process of fighting a cancer, the tumor should be a logical site for isolating large numbers of T cells on the attack. Removing a small piece of tumor from a mouse, they cultured the cells with IL-2 for several weeks. The culture was overrun with voracious T cells dubbed *tumor infiltrating lymphocytes* (TILs). Injecting these newly isolated cells into cancerous mice resulted in a marked regression of their widespread tumors.

Soon human volunteers became the source of TILs. The *New England Journal of Medicine* reported in December 1988 that 11 out of 20 patients who had advanced melanoma showed substantial regression of their tumors after they each received IL-2 and about 200 billion TILs grown from their tumors. But were there among the billions of TILs given to each patient some cells which were much more effective than others? Why were some patients helped and not others? And how did the cells circulate in the blood and home in on the tumor site? A way to label the cells was needed to follow their fate.

Rosenberg proposed to employ a mouse leukemia virus, in a form harmless to humans, in order to insert a gene into the chromosomes of the TILs. The latter had been cultured from a golf-ball-sized piece of a tumor removed from each patient. The gene would confer on these cells the ability to make a protein which would render them resistant to the antibiotic neomycin, a substance which normally kills human cells. Each patient would then be transfused with his or her specific modified TILs. Then, as these TILs wended their way throughout the patients' bodies, the transformed cells could be identified by taking blood and tissue samples and exposing them to neomycin. The gene-altered cells would be the only survivors. All of this, of course, was contingent on the ability of the introduced genes to function in the body in a detectable way.

The experiment, begun on May 22, 1989, worked beautifully. Dr. Rosenberg (and 14 co-authors) reported in the August 30, 1990, *New England Journal of Medicine* that the genetically transformed TILs could be traced in the patient's circulation for as long as two months and that these cells were found in the tumors as much as 64 days after cell administration. They had demonstrated that their retrovirus could safely transfer genes into human lymphocytes, which would then express those genes after being transferred to a human subject. The first giant step toward gene therapy had been taken.

The rocky road to full approval for this first human gene transfer had included more than pointed queries from the review panels. At the January 30 meeting of the Recombinant DNA Advisory Committee, Jeremy Rifkin launched an attack to block the experiment and all human gene therapy research until NIH agreed to set up an "Advisory Committee on Human Eugenics" to evaluate the ethical and social implications of such studies. He further filed a lawsuit in federal court to halt the experiment, based on his contention that the public was not present for the final review.

NIH officials answered that the mail ballot asking for an official vote was a courtesy to committee members who had earlier indicated their approval. A court settlement instructed the committee to rewrite its charter to ensure that decisions would be made in full consultation with the public. Rifkin, who claims that he never intended to block the experiment but only assure that there was sufficient public input, was not granted his request for a "eugenics" committee.

Eugenics, defined as "methods of improving the quality of the human race, especially by selective breeding," conjures up images of past abuses—forced sterilization, harsh immigration laws, and the genocide practiced by Germany under the Third Reich. While these shadows of the past always lurk in the background of debates about genetic science, as well they should, NIH let it be known that it had no intention of establishing a redundant committee with such an inflammatory title.

The medical potential of genetically altered lymphocytes extends

far beyond the initial gene transfer that we have just described. Researchers soon broached the idea of using such cells as suitable vehicles
for introducing genes to treat a wide range of other diseases. Many
hitherto incurable diseases waited in the wings for a genetic solution,
such as other forms of cancer, hemophilia, SCID, cystic fibrosis, and
muscular dystrophy.

Now, only two years after the world's first attempt to cure an illness
with gene therapy, events are unfolding rapidly. At this writing 40 gene-
transfer clinical trials with humans have already been approved worldwide, 18 of which are specifically for the purposes of gene therapy. The
others are designed to trace the fate of genes introduced into the body,
with the hope that this information may ultimately lead to a therapeutic
procedure. Numerous other gene transfer protocols are at various levels
of preapproval development.

We will look at what may be the most promising of these shortly.
First, let's look at the scientific principles and technological approaches
common to many of the procedures of gene therapy. In understanding
those in the context of their development we will establish a frame of
reference for what promises to be a revolution in genetic medicine. Let's
do that by going behind the scenes to better understand how the first blow
was struck against our genetic fate.

THE FIRST STEP

Unquestionably, a central figure in the unfolding drama of gene
therapy has been the seemingly ubiquitous W. French Anderson. The
dream of understanding genetic disease at a molecular level had captured
his imagination early. He wrote about those dreams on his application to
Harvard in 1953, that pivotal year in which James Watson, Francis Crick,
and colleagues described the structure of DNA. Four years later, in
Anderson's senior year at Harvard, Watson would be his instructor in one
of the first courses linking genetics and DNA.

The very next year was spent with Francis Crick at the Cavendish
Laboratory in Cambridge, England, where he assisted in Crick's effort to

prove that the genetic code is organized triplets of nitrogenous bases. Completing his M.D. at Harvard in 1963, he went on two years later to NIH, where he worked for 27 years. In 1992 he moved to the University of Southern California's Norris Cancer Center, where he set up a gene therapy research institute.

In 1966, newly arrived at NIH, he submitted a paper titled "Current Potential for Modification of Genetic Defects" to *The New England Journal of Medicine*. In it he envisioned viruses carrying genes into cells. Despite being recognized by some reviewers as "very erudite and fascinating" it was rejected as being "too speculative."

At NIH, Anderson began working as a postdoctoral student under Marshall Nirenberg, who was working out the final pieces of the puzzle of the genetic code. Anderson first achieved prominence of his own when in 1970 he discovered specific factors in cells which govern the synthesis of hemoglobin. A year later he applied this information to reproduce the molecular defects caused by two hereditary diseases, thalassemia and sickle cell anemia.

Anderson had already made it quite clear to his NIH colleagues that he intended to devote himself to unraveling the mysteries of human genetic diseases. Such was the rudimentary state of knowledge of the cellular details of human genetics only twenty years ago that it hardly seemed a promising career path for a bright young man.

He turned his attention first to thalassemia, the inherited condition in which children are unable to make normal hemoglobin, the complex red blood cell molecule which transports oxygen. Deprived of healthy hemoglobin, a child with severe thalassemia bears the burdens of severe anemia, a grossly enlarged spleen, and bone deformities. Anderson had become an expert on the intricacies of the hemoglobin molecule, but as the 70s gave way to the 80s he despaired of confronting the genes responsible for its altered synthesis. As it turned out, the instructions for producing hemoglobin are encoded in not one but a complex of genes which interact to precisely balance the amount of hemoglobin in red blood cells.

What was needed for the first gene-treatment candidate was a disease caused by a single gene. If its defective protein product were replaced by the activity of a normal gene introduced into the patient, the products of the introduced gene should be effective and safe in a wide range of concentrations. Ideally the defect would be found in the bone marrow, since some techniques were already in place for marrow manipulation. Finally, of course, the normal gene would have to be isolated and copied so that the researchers would have an abundant supply.

In *Science* (October 1984) Anderson described a prime candidate. It was adenosine deaminase deficiency (ADA), an inherited lack of a crucial cellular enzyme. (We identified ADA earlier in this chapter as a form of SCID.) Approximately 35 percent of SCID cases (there are only about 50 children known to have SCID) are due to the lack of the gene which directs the production of adenosine deaminase. This seemingly simple defect results in a profound decrease in the numbers of the main players in the body's defense, the crucially important T cells. Loss of T cells causes a decrease in the number of B cells, another form of lymphocyte which makes antibodies. Most infants with ADA die soon after birth from overwhelming infection.

At the time of Anderson's research, the normal ADA gene, known by mapping studies to be on chromosome 20, had already been cloned. Moreover, it was known that humans could tolerate a wide range of ADA enzyme activity. Individuals with enzyme levels from as low as 5 percent of normal to 40 times the normal levels have normal immune function. This was a great advantage, because the level of enzyme production from genes that might be inserted into a person would be unpredictable. Finally, ADA was also known to be a disease which could sometimes be treated successfully with bone marrow transplantation, although fewer than 35 percent of people with ADA have a compatible donor available.

After carefully reviewing all of the possible ways that one might deliver normal genes to someone with ADA, Anderson settled upon the following scheme. Bone marrow would be removed from the person and treated with retroviruses containing the normal gene. Some of these would infect marrow cells and introduce the gene into their genome. When the bone marrow was put back, the genes would function, producing the normal ADA enzyme, and a healthy immune system would develop.

Anderson called for a thorough scrutiny by review boards to evaluate the entire procedure. He particularly emphasized that safeguards be placed against the possibility of harm being done by the accidental introduction of malignant cells or harmful viruses. He recommended that gene activity first be demonstrated in lab bone marrow cultures and in animals. Finally, while emphasizing that it would be unrealistic to expect a complete cure from the first trials, if stringent criteria of safety and effectiveness were met he thought that "it would be unethical to delay human trials" since there was no other reasonable hope of a cure.

It would be six years before the same trio of Anderson, Blaese, and Rosenberg, who had already performed the first gene-transfer procedure, described earlier, would be given permission to administer genes for treatment. The road to approval, which finally came after three years of line-by-line scrutiny, had almost been blocked earlier that year by the announcement of a new drug for ADA treatment. Injections of the natural ADA enzyme had never been effective because it remained in the blood for only a few minutes. However, if the enzyme is coated with PEG (polyethylene glycol, the basic ingredient in antifreeze), it would survive in the blood for days and assist the child in fighting off infections. But at $250,000 per year for weekly injections it was still not a practical treatment.

The researchers accepted a compromise—the children to be treated with genes would also continue to receive this medication, even though the first patient, the little four-year-old girl, had not responded well to PEG–ADA after some initial improvement. Evidence was still needed to convince reviewers that gene therapy would be an improvement over PEG–ADA. The answer came from Italy. Claudio Bourdignon flew to Washington from Milan to report to the reviewers. He told them that human lymphocytes that had been given ADA genes in viruses showed normal immune function for up to three months. Final approval followed on June 30, 1990.

RETROVIRUS MESSENGERS

Viruses are not living organisms. So small that they cannot be seen even with a powerful light microscope, they are merely nucleic acid, either DNA or RNA, wrapped in a bit of protein. They can propagate themselves only by entering a living cell and taking over the metabolic machinery of the cell, forcing it to make viral copies. They are strange, lifeless parasites which emerge and pass from cell to cell, eventually degenerating if they cannot find and enter another cell and force it to do their bidding.

A retrovirus is one which contains two identical strands of RNA rather than DNA. Probably the most conspicuous example of a retrovirus is the human immunodeficiency virus (HIV), which attacks T cells. The wholesale destruction of T cells results in AIDS. The protein coat surrounding the retrovirus is covered by a membranous envelope. This membrane enables the virus to recognize unique areas, the *receptors*, on the surface of specific cell types. In a viral attack, after the virus attaches to a receptor site, the virus enters the cell and sheds its envelope. Then, in a step unique to retroviruses, it releases an enzyme, *reverse transcriptase*, which makes a DNA copy of the viral RNA using building blocks present in the host cell.

The DNA enters the nucleus and is integrated into a host cell's chromosome. It begins to function as part of the host genome and is now referred to as a *provirus*. It is a messenger which has arrived in the nucleus bearing one simple message for the cell—make more viruses. The cell is obliged to comply.

Gene therapy using retroviruses takes advantage of this extraordinary capacity of retroviruses to enter a cell's genome and become an active part of it (see illustration). Using recombinant DNA technology, which we discussed in Chapter 1, researchers remove several retroviral genes from a suspension of retroviruses and replace them with (for example) ADA genes. Certain genes are cut out of the viruses to render them harmless to the human recipient.

"Helper" cells then are employed to make large numbers of viruses which initially lack RNA. These helper cells are then bathed in a suspension of ADA genes complexed with another small bit of DNA and are chemically treated to stimulate uptake of DNA. Some of the ADA genes enter the helper cell's genome and are copied in the form of RNA. The DNA sequence that has arrived accompanying the ADA gene enables the RNA to be packaged inside viral proteins, and the newly engineered

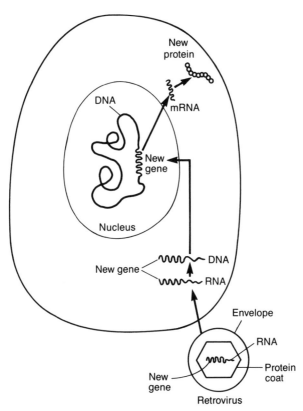

The new gene is put into the RNA of a *retrovirus*. When the retrovirus injects its RNA into the cell, the new gene is carried along. The DNA made by *reverse transcriptase*, along with the introduced new gene, may become part of a chromosome in the nucleus. This gives the cell the ability to make whatever proteins are coded for in this introduced genetic material. This illustration is a simplified summary version of what is actually a more involved procedure.

viruses, now carrying ADA gene messages, emerge from the helper cells and are collected.

The original protocol was next to infect bone marrow with these viruses in order to transfer the ADA gene message to blood-forming marrow cells. Instead, using some of the techniques already proven successful in the first gene-transfer protocol, already described, a concentrated sample of the young patient's white blood cells was prepared and cultured to raise a dense population of T cells. These were then incubated with the engineered viruses, which also carried the neomycin resistance gene so the cells could be followed.

Over the next 10 months the little girl received seven such intravenous infusions, together with the weekly PEG–ADA injections. So successful was this regime that gene-corrected T cells made up almost 20 percent of her circulating blood cells. No ADA treatment was given for $6\frac{1}{2}$ months, after which she was put on a schedule of maintenance infusions at 3- to 5-month intervals.

The "Second National Conference on Genetics, Religion, and Ethics" convened at the Texas Medical Center in Houston on March 13–15, 1992. Those of us who were in the audience in the opening session heard the evident pride in the voice of French Anderson as he described how his young patient was no longer constantly ill. He related how, when her whole family came down with the flu, she was the first to recover. She was living the life of a happy kindergarten student, never having shown any serious side effects. Another patient, a nine-year-old girl with ADA who had begun an identical course of treatment in January 1991 was also doing well.

In November 1992 the two girls, now six and 11, were the guests of honor at a party held at the Smithsonian Institution to celebrate the second anniversary of the world's first attempt to cure a genetic disease with gene therapy.

Is this a permanent genetic cure? Technically, no, not as long as treatments must continue. T cells may survive in the bloodstream for months or even years—but not for a lifetime. Ideally, one would want to alter cells which could act as a source of the missing enzyme indefinitely. Impossible? Not necessarily—for such cells do exist. They are the long-sought-after *stem cells*, which are found in very small numbers hidden in the complex mixture of the bone marrow.

The Messenger in the Marrow

Carried along in the pulsating current of human blood is a remarkable variety of cells, each precisely tailored to specific functions. White blood cells, which include the lymphocytes, monocytes, and neutrophils, among others, defend against attack by foreign invaders such as bacteria or viruses. The red blood cells carry vital oxygen to all the body's tissues. Platelets, not actually cells but fragments of large cells from the bone marrow, can ward off excessive bleeding by initiating clotting.

The life span of blood cells varies widely. Red cells survive for about 120 days, while platelets must be replenished within one week. Some lymphocytes, on the other hand, may persist for years. Amazingly, all of this complex array of cells develops from a single source, the stem cells, most of which reside in the bone marrow, where they account for only about one in every 10,000 to 100,000 cells. The stem cells begin their work very early in an individual's life. Within a few weeks after conception, they have formed in the embryonic yolk sac. From there they later migrate to the liver and then to the bone marrow, where blood cell production is in full swing by the eighth month of fetal life. At birth, active marrow is in almost every bone, turning out new blood cells at the rate of several billion per hour. In the adult, blood cell manufacture is limited to the ends of long bones. Meanwhile, there is always a small reservoir of stem cells in the marrow and circulating blood, ready to develop further in differentiated blood cells.

Bone marrow transplants have proven useful for at least temporary relief in SCID and other serious blood cell disorders such as certain anemias and leukemias, as well as after chemotherapy or radiation therapy. The ideal donor in these cases is a sibling whose marrow cells share so many genes with the recipient that they are not rejected, because they are not recognized as foreign. Otherwise, unrelated donors may be used if they meet a certain level of compatibility.

After local anesthesia the rich, fluid marrow is slowly withdrawn through a needle inserted into the marrow cavity within the bone, usually a hip bone in adults or a leg bone in young children. After whatever treatment to the marrow is necessary, the marrow is gradually returned via the patient's veins. Ferried by the bloodstream, the marrow cells find their way back to the bones and take up residence once again in the marrow cavities.

If one could selectively remove the totipotent stem cells (so called because they can produce all blood cells) and introduce new and useful genes into their genome, in theory at least, there would no longer be a need for lifelong injections of short-lived genetically transformed cells. The stem cells, as they developed into mature blood cell types, would pass on the introduced genes. One problem with this ideal scenario is that stem cells have been notoriously difficult to isolate and grow.

On February 11, 1992, the Recombinant DNA Advisory Committee announced that it would give the team of Blaese, Anderson, and colleagues permission to add a new technique to its ADA protocols. When they had begun their patient trials in September, 1990, there was no simple way to target the stem cells for genetic transformation. But in the meanwhile, several research teams and biotechnology companies had made remarkable progress by using an antibody that ferreted these cells out of a complex mixture of other cell types because it would attach to a site on the cell membrane called the CD34 site. The procedure was put on the market by CellPro, Inc., a biotech firm in Bothell, Washington.

The NIH group proposed to use the CD34 technique, not by using bone marrow but by pooling white cells from blood samples and removing CD34-positive cells. These would be treated with the normal ADA-bearing retrovirus and transfused back to the young patients. As they waited final clearance from the FDA and the NIH director, the Milan, Italy research group led by Claudio Bourdignon announced that, after approval by the Italian Committee for Biosafety, it had treated a 5-year-old child for ADA by administering gene-modified cells rich in stem cells collected from the child's blood.

By the time the first International Workshop on Human Gene Transfer had convened in Chantilly, France, on April 11, 1992, mouse stem cells had been purified. SyStemix, Inc., a medical research firm in Palo Alto, California, had announced that it had received a patent on its stem cells, its method for isolating them, and uses of the cells in medical therapy.

Perhaps stem cell therapy will become routine. Intriguing possibilities could arise. Perhaps as researchers learn to cultivate certain stem cell lines these could be induced to become effective immune cells. A test tube full of B cells might be induced to learn how to make antibodies against infectious agents such as the AIDS virus or cancer cells. Healthy people might arrange to have a sample of their stem cells put into cold storage until the cells might be needed to treat a serious disease.

We have only just begun to use genes in our battle against cancer and other human diseases. Bear in mind, human gene therapy is still in its very early stages. Given the rapid rate of progress and the creativity of scientists, any chapter on the subject must necessarily lag behind the latest medical headlines. However we can anticipate what may develop over the next decade by briefly reviewing clues that scientists already have which may lead to tantalizing possibilities.

Genes versus Cancer

Experts predict that the major emphasis in human gene therapy experimentation over the next few years will be directed at the treatment of cancer. Emphasis will be on genetically modifying cells of the body's own immune system to fight tumors. In this way, the same genes may be used to treat many different forms of cancer.

On July 30, 1990, at the very same RAC meeting at which the Blaese–Anderson proposal for ADA gene therapy was approved, the board gave the go-ahead to a landmark study in the fight against cancer. There is a pressing need for innovative treatments against this devastating disease. Each year more Americans succumb to cancer than died in World War II and the Vietnam War combined. The new proposal was initiated by Steven Rosenberg as the head of a team which again included French Anderson and Michael Blaese.

The target was metastatic melanoma, a skin cancer that is particularly difficult to treat and whose incidence is increasing by 4 to 5 percent each year. The new experimental approach was based on the information gleaned from the historic human gene-transfer experiment described earlier in this chapter, in which tumor-infiltrating lymphocytes (TIL) marked with an antibiotic-resistance gene were detected after being injected into patients with melanoma.

The new weapon was a gene which controlled the production of a powerful naturally occurring antitumor protein known as *tumor necrosis factor* (TNF). Rosenberg had remodeled the gene so that it could generate up to 100 times the normal concentration of TNF. Since they had already shown that at least some injected TIL cells migrate to the tumor site, it seemed a logical step to use a retrovirus to introduce their TNF gene into TIL cells which could deliver them to the tumor. There it could generate TNF and destroy the tumor.

In 1992 the scientists began another modification. Now, either the TNF gene or the gene for interleukin-2, the immune-cell growth stimulant, was inserted into tumor cells that had been isolated from patients and grown in culture. After this gene modification, the tumor cells were injected beneath the skin of the patients in hopes of enhancing their immune systems' ability to recognize and kill all remaining tumor cells.

By late 1992 several important aspects of the TNF–TIL experiments had not yet worked out as the researchers had hoped. The injected TIL cells were moving to the tumor sites, but many ended up elsewhere in the body, and the modified cells were not making TNF at levels high enough to affect the tumors.

As these clinical trials were proceeding, the menace of melanoma was attacked on a different front. In June 1992, University of Michigan molecular geneticist Gary J. Nabel announced that "We have begun to use DNA as a drug." His approach was to imitate the rejection that occurs in organ transplants. In such rejection the recipient recognizes certain proteins, the transplantation antigens, that occur on the surface of cells in the new organ, causing the immune system to attack it.

Since most cancer cells do not carry such antigens, Nabel put genes to work to put them there. He and his colleagues took the gene for a transplantation antigen, HLA-B7, and, eliminating the need for retro-viruses, instead simply encapsulated the genes in tiny fat globules termed *liposomes*. These, when injected into tumors in laboratory animals, had bound to the tumor cells and carried the genes into their interior. The genes entered the genome and began to churn out HLA-B7, which then moved to the cell membrane, where it was recognized and attacked by the immune system. A 67-year-old woman was the first of 12 patients to try this novel therapy. These early trials were designed primarily to test the safety of the technique and to determine the proper dosage.

In mid-1992, Scott M. Freeman and his colleagues at the Rochester, N.Y., Medical Center were given final approval for yet another gene methodology, this time in the fight against ovarian cancer. Tumor speci-mens would first be removed from the patients and grown in the laboratory. Using a modified herpes virus, they planned on incorporating into the tumor cells a gene which codes for an enzyme that plays a central role in DNA synthesis and is sensitive to the antiviral drug gancyclovir. They had found that after this procedure was carried out with mice, gancyclovir killed not only the newly added cells but other cancer cells as well.

At about the same time, Kenneth Culver, of Gene Therapy, Inc., in Gaithersburg, Maryland, announced that he would use a similar ap-proach to launch an attack on brain tumors. He had already shown that he could cure 80 percent of certain kinds of brain tumors in mice with gene injections followed by gancyclovir treatment.

As 1992 grew to a close, the International Conference on Gene Therapy of Cancer was held in San Diego, California. The proceedings revealed a remarkable burst of activity in research on gene-therapeutic techniques, as well as reports on success in identifying genes that either inhibit cells' ability to mutate into tumor cells or cause tumor cells to revert to normal ones. By early 1993 over a dozen clinical trials had been approved for gene therapy aimed directly at the cure of various cancers, including ovarian, melanoma, brain, kidney, and lung cancers.

Despite the fact that there are still many technical problems to overcome before gene therapy for cancer can become standard treatment, if research and testing continue over the next few years at the current rapid rate, the decade of the 90s may perhaps see genes become a powerful tool in the fight against cancer, one of humanity's most formidable enemies.

Cystic Fibrosis

Cystic fibrosis is among the most commonly inherited disorders among Caucasians. There are now about 30,000 people with CF in the United States. Children who inherit two copies of the faulty gene, one from each parent, are not able to excrete chloride from the moist cells lining the lungs, sweat glands, intestines, and pancreas. Thick mucus builds up, particularly in the lungs, breeding numerous infections. Few survive beyond their twenties.

In a triumph of molecular biology, Francis Collins, Lap-Chee Tsui, Jack Riordan, and their co-workers ended a 10-year quest in August 1989. They had tracked the CF gene to its home on chromosome 7. Soon several groups had managed to put the gene into lung cells isolated from CF patients, resulting in the restoration of normal chloride transport across the cell membranes. Then Ronald Crystal at NIH injected the lungs of rats with a nasal drip of adenoviruses (cold-causing viruses) that had been stripped of their infective genes and replaced with normal genes. The rat lung cells produced the normal human gene product, CFTR, for up to six weeks.

In December 1992 the NIH approved three proposals to use genetically altered cold viruses into the lungs of 25 people suffering from cystic fibrosis. The experiments were designed to test the safety of this technique, rather than its effectiveness as a treatment. The trials were to be carried out at the National Heart, Lung, and Blood Institute in Bethesda, Maryland, the University of Michigan in Ann Arbor (site of the research headquarters for the Cystic Fibrosis Foundation), and at Iowa University, in conjunction with the Genzyme Corporation. Genzyme, based in Cambridge, Massachusetts, has already committed more than $300 million to find a cure for cystic fibrosis.

These trials are the first time that adenoviruses have been employed for introducing genes into humans. While gene therapy trials to combat cystic fibrosis are just getting under way, officials at the Cystic Fibrosis Foundation are expressing cautious optimism for the future of CF gene-based treatments for this devastating disease.

Muscular Dystrophy

One of every 3,500 boys is born without a functional gene for producing the important muscle protein dystrophin. This deficiency leads to muscular dystrophy, a condition in which the muscles gradually waste away. Boys born with Duchenne muscular dystrophy, the most common childhood form of the disorder, usually die in their early 20s. There is no known cure.

Unlike ADA, in which genes can be integrated into blood cells which are then returned to the patient, in MD the genetic defect is seemingly inaccessible, hiding deep within the long, fibrous cells of the skeletal muscles. Nevertheless, in 1991 Jon A. Wolff of the University of Wisconsin–Madison reported that he had directed a fruitful study using technology developed at Vical, Inc., in San Diego. The thigh muscles of mice deficient in dystrophin had been injected with copies of the normal dystrophin gene linked to plasmids, those small circular DNA molecules from bacteria used in recombinant DNA manipulations. The very large gene is too big to fit into retroviruses, and so the plasmids were used as a vector. Unfortunately, only one percent of the muscles of the mice took up the new gene and manufactured dystrophin. Much more dystrophin would be needed for healthy muscle function, but these results were convincing evidence that the genome of muscle cells could be reached and affected.

In 1990 Peter Law at the University of Tennessee in Memphis had pioneered an ingenious approach to a potential MD treatment. He injected the big toes of several young MD patients with immature muscle cells taken from either their fathers or their brothers. The small fragments of donated muscle had been first cultured in the laboratory. Within days, millions of small, elongated myoblasts—repair cells that can rebuild injured muscle—grew from the small reservoir of myoblasts always present. Those myoblasts, when injected into the toe muscles, fused with them and began to produce some dystrophin.

Later in that same year, two research teams, one led by Eliav Barr and Jeffrey Leiden of the Howard Hughes Medical Institute and the other by Helen Blau at the Stanford University School of Medicine took advantage of myoblasts in developing another ingenious way to ferry genes into muscles. They put the gene for human growth hormone into myoblasts and injected these into mouse muscles, resulting in significant blood levels of the hormone for up to three months.

Will the future bring news that the dystrophin gene can be insinuated into muscles by myoblasts or injection or other means in enough quantity so that once-useless muscles can function again? Or is more than just dystrophin needed? Researchers predict that myoblast gene therapy may have an equally promising future as a delivery method for many other therapeutic genes. Given methods for controlling the rate of activity of newly introduced genes, not an easy challenge, we will perhaps be able to fashion myoblasts which when implanted will not only create dystrophin where needed but release insulin to control diabetes or generate Factor VIII, the blood-clotting protein missing in one type of hemophilia.

The Heart of the Matter

The cholesterol in our blood is a mixed blessing. It is a vital chemical which makes up part of the membrane of every cell in the body. Unregulated, it can lead to an early death. The concentration of cholesterol in the blood is the result of a balance struck by our dietary intake and cellular metabolism. A crucial element of that control is the LDL receptor, a protein which pokes through the surface of liver cells and snares passing molecules of low density lipoprotein (LDL), often referred to as the "bad" form of cholesterol. Transported inside the cell, the LDL is digested and removed.

One in 500 people inherit one abnormal gene for making LDL receptors, causing a mild form of familial hypercholesterolemia, while 1 out of a million are born with a severe case of this genetic disease. In either situation, blood cholesterol levels are high, in the latter instance so high that the person develops severe coronary artery disease and rarely survives beyond the second decade of life.

In a few cases people have been helped by the drastic procedure of a liver transplant. Now, there may be a way to bolster the LDL filtering capacity of the genetically deficient liver through gene therapy. James Wilson and co-workers at the University of Michigan Medical Center have used retroviruses to place the normal LDL receptor gene into liver cells removed from rabbits with hereditary high cholesterol. Transplanting the genetically corrected cells into the donor rabbits' livers resulted in a 30–40 percent decrease in their blood cholesterol.

Further, Wilson has been able to likewise genetically modify human liver cells. In the first gene therapy protocol approved to be conducted outside of the Clinical Center of NIH, Wilson's team delivered LDL genes to human volunteers. After the first patient to receive the treatment experienced a 15 to 25 percent drop in blood levels of LDL, Wilson and his colleagues were given permission to conduct their gene therapy experiments on five more people who had inherited the tendency to develop dangerously high levels of cholesterol.

An equally intriguing possibility for the future treatment of cardiovascular disease lies in research on implanting genes into the endothelium, the thin, smooth layer of cells which lines our blood vessels. Active genes put into the endothelium would be in a perfect position to secrete useful proteins directly into the vascular system. Progress has been made toward this goal. Several research groups have already been able to insert genes into pig endothelial cells growing in laboratory culture. In some instances these cells could be put back into pig arteries, where they then functioned normally.

Others have managed to get functional genes enclosed in liposomes into endothelial cells in dogs and have placed genes in animal arteries by implanting vascular grafts seeded with genetically modified cells. In the future we may see such grafts used to deliver a wide range of therapeutic genes, such as the gene for tPA, a naturally occurring chemical which aids in dissolving blood clots. Many other conditions needing gene replacement might be targeted as well besides those that are implicated in cardiovascular disease. As usual, a major stumbling block will be to develop gene constructs which will put enough gene product into the blood at safe, useful concentrations.

AIDS

Paradoxically, gene therapy researchers are calling on the deadly AIDS virus HIV to carry resistance into T cells, the very cells in the human immune system which the virus targets for destruction. The scientists first remove the infectious genes (in the form of RNA in the retroviral HIV) and replace them with genes designed to block HIV replication. Trials will soon be under way in which blood will be taken from HIV-infected patients. T cells that remain capable of identifying and killing other cells harboring the virus will be carefully weeded out of the sample and cultured in large numbers. The hope is that these, when returned to the patient, will seek out and destroy infected cells.

Other ways of tricking infected cells into harming the virus are in the planning stages. We may someday be able to target genes to infected cells where the genes would trigger cell suicide. Another possibility might be to stimulate many cells to make CD4, the cell surface protein recognized by the AIDS virus, thus luring it to attach itself to the decoy CD4 rather than that of real T cells. This would spare the vital T cells from destruction and leave the virus clinging to CD4 and thus unable to do further harm. French Anderson is collaborating with Dr. Robert C. Gallo of the National Cancer Institute on just such an effort.

One biotechnology company, Viagene of San Diego, California, is getting set to carry out its approach to AIDS gene therapy. Fibroblasts, cells from deep within the tissues supporting the skin, will be removed from patients' arms and genetically changed to make a protein that is part of the outer coating of the AIDS virus. These will be put back into the patients, in the hope that the modified cells will activate the body's immune system to destroy any cell carrying this protein on its surface— including those cells actually infected with the AIDS virus.

In the short run, making better use of existing drugs and searching for new ones will continue to be a major priority. But in the background, gene therapy remains an ideal whose time may come.

THE GENE BUSINESS

According to French Anderson, gene therapy will become a major weapon in the fight against disease only when vectors are developed "that can safely and efficiently be injected directly into patients as drugs like insulin are now." We will need to engineer those gene-delivery systems so that they "will target specific cell types, insert their genetic information into a safe site in the genome, and be regulated by normal physiological signals." Any success in gene therapy will of course bring with it the possibility of profit for those who have proprietary rights to its use.

In the wake of preliminary successes, companies devoted to gene therapy have sprung up and others have joined in the hunt. Research scientists do not always work exclusively in an academic, industrial, or governmental setting. Often they combine research in some or all of these venues. Partnerships between private corporations and government scientists have been officially encouraged since 1986, when Congress passed the Technology Transfer Act to facilitate the marketing of government inventions. Congress felt that the development of many useful discoveries was being slowed simply because no one could make a profit from a government-held patent.

Their answer was CRADA, the Cooperative Research and Development Agreement, which permits companies to hold licensing rights to patents held by government agencies. Under a CRADA, the government scientists can be held to a maximum royalty of $100,000 per year from private collaborative agreements. Beyond the profits, such arrangements are often advantageous because of the facilities and staff which the scientist has at his or her disposal. French Anderson, for example, was the first NIH scientist to sign a CRADA, for his work with Genetic Therapy, Inc., in Gaithersburg, Maryland. Working with GTI, which had developed the retrovirus for Anderson's ADA therapy, afforded him far more resources and personnel than NIH could provide.

Currently, a number of companies are active in gene therapy research. In 1991 the three original gene therapy biotechnology companies expanded. Viagene, Inc., in San Diego, sold worldwide marketing rights to an AIDS therapy it was developing to the Green Cross Corp. of Japan. GTI went public and began a collaboration with Sandoz Pharmaceuticals, Inc. Somatix Corp. merged with Hana Biologicals and then with GenSys to become Somatix Therapy Corp. By 1992 seven companies were gene therapy specialists, and other biotech companies were adding this specialty to their repertoire.

To the winner(s) belong the spoils. More important, to those now without the blessing of a cure for their genetic disease, the prospect of some of our best scientific minds searching for just such a cure with the support of abundant resources, whether private or public, offers a ray of hope.

THE ETHICAL CHALLENGE

In the early pages of this chapter we defined four categories of genetic manipulation: somatic-cell therapy, germ-line therapy, enhancement genetic engineering, and eugenic genetic engineering. As we have subsequently shown, so far only a few somatic-cell therapy procedures have actually been tried with humans, although many more will soon be done. We are far from having the technological capability to experiment in any of the other three categories, at least with humans.

In 1980 French Anderson, along with ethicist John C. Fletcher, pointed out in an article in *The New England Journal of Medicine* that gene therapy would be possible in the near future. They warned that scrupulous care should be taken to determine and meet the conditions under which such a procedure would be ethical. They took the position that the illness to be treated would have to be so severe that even a considerable risk would be justified.

In addition, they stressed that animal studies should first be completed in which newly introduced genes should go to the correct cells and remain there—or if they did leave those cells and go elsewhere in the body, that they should do no harm. Next the gene product should be regulated so that it is not released in harmful amounts. Finally, the inserted genes should not harm the genome. The latter point is an important one. You may recall that when a gene from an external source is added to a genome by means of, for example, retroviruses or liposomes, the integration of that gene into the genome is completely undirected.

In other words, there is at present no way to predict exactly where the new introduced gene will find a home on one of the chromosomes of the recipient cell. This means that there is a very slight but real chance that the new gene might end up within a gene already there. The function of that resident gene might be vital to the function of that cell and its interruption could prevent its normal functioning.

Should that happen, one possibility is that this one cell could die, in which case the negative effect would be negligible. But one could imagine a situation in which the new gene might end up where it would activate a previously dormant oncogene, one of the cancer-causing genes which we all have in our cells. Or it could conceivably inactivate a tumor supressor gene, busy making proteins which the body uses to destroy cancerous cells. Others in the scientific community echoed cautions similar to those of Anderson.

Actually, the 1980s and 90s have been a time of unprecedented public debate over genetic manipulation, and scientists have been quite as vocal as ethicists, philosophers, theologians, and other participants. A general consensus has arisen that somatic-cell gene therapy used for treating a serious disease is an ethical option. An analogy is often made between it and organ transplant, the major difference being that genes are being transplanted rather than an entire organ. Somatic-cell gene therapy is considered to be supported by the basic moral principle of beneficence because it would relieve human suffering.

Leroy Walters, Director for the Center for Bioethics of the Kennedy Institute of Ethics at Georgetown University, has identified 20 statements on human gene manipulation formulated between 1980 and 1990. The sources included committees in Denmark, Sweden, and Australia, as well as the Parliamentary Assembly of the Council of Europe, the World Medical Association, the Catholic Church, and the World Council of Churches. All 20 accepted the moral legitimacy of somatic-cell gene therapy to ameliorate serious human disease. However, the majority opposed germ-line gene intervention, and none supported the enhancement of human capabilities by genetic means.

Why the approval of the first and not the others? Well, they realize that germ-line gene therapy would alter egg cells, sperm cells, or early embryonic cells. Because the added genes would pass from cell to cell as prenatal development continued, the genes would end up in all or many of the body cells, including the eggs or sperm. The changes could thereby affect not just the individual developing from the modified cells but all or some of his or her descendants.

When germ-line genetic alterations are envisioned for therapeutic purposes, one usually thinks of a scenario in which egg and sperm have been joined together outside of the body, literally in a laboratory dish in the now rather common process of *in vitro* fertilization. Following the fertilization of several eggs, a cell can be removed from each of the young few-celled embryos, and the cellular DNA can be amplified by PCR and tested for genetic defects. This is termed *preimplantation diagnosis*. It is of course highly technical and very expensive. In the few instances where this is done, the physician has the option of selecting only those embryos free from defects for implantation into the uterus. Germ-line gene therapy would offer the option of repairing the genetic defect.

It is easy to see that while these highly technical manipulations might be used to prevent certain diseases and to remove the possibility of certain undesirable genes from being passed on to future generations, it is obvious that much the same method could be employed to add "good" genes to "enhance" rather than cure. For example, if genes were identified and isolated which would endow a person with an excellent immune system, which might prevent him or her from ever suffering from an infectious disease, it might be tempting to want to endow future humans with such an enhanced defense system.

This "slippery slope" argument, that once under way germ-line gene therapy would inevitably lead to enhancement and eugenic genetic engineering, along with the fact that germ-line alteration is really at the drawing-board stage, has led to a tendency to dismiss it out of hand. In 1988 the European Research Council asserted that "Germ-line gene therapy should not be contemplated." In 1990 the Recombinant DNA Advisory Committee simply announced that "the RAC and its Subcommittee will not at present entertain proposals for germ-line alterations."

Other objections have been raised. For example, some have argued that it would violate the rights of generations not yet born to force them to inherit an intentionally modified genome. Others stress the unacceptability of the long-term risks, or point out that it would force future generations to be unconsenting research subjects.

In the December 1991 *Journal of Medicine and Philosophy*, an issue devoted to the theme of "Germ-Line Gene Therapy," geneticist Edward M. Berger and philosopher Bernard M. Gert take the position that until we have almost certain knowledge about the risks involved, we cannot justify trying to confer a benefit on a relatively few people because there would be an unlimited number of people put at risk in the future.

Whatever the arguments, there are signs that the pendulum has shifted away from almost automatic rejection toward the feeling that the promises and perils of human germ-line gene therapy ought to be examined in a systematic and open way. For instance, the summary statement formulated by the participants in the Council for International Organizations of Medical Sciences' conference on "Genetics, Ethics and Human Values: Human Genome Mapping, Genetic Screening, and Therapy" held in Tokyo and Inuyama City, Japan, in July 1990, offers the following position. It states that "although germ-cell gene therapy is not contemplated at present, continued discussion of germ-cell gene therapy is nonetheless important. The option of germ-cell therapy must not be prematurely foreclosed. It may someday offer clinical benefits attainable in no other way."

In 1991 Leroy Walters called for "forums, or even a sustained study process, regarding germ-line genetic intervention." A year earlier, ethicist John C. Fletcher had written that "to argue that it should never be done at all is unreasonable at the least, examined from the standpoint of the well-being of future generations."

Even French Anderson, who is adamant in his opposition to any form of enhancement or eugenic genetic engineering, strongly supports germ-line gene therapy, but only if it meets strict criteria:

1. it must be preceded by years of experience with somatic-cell therapy,
2. there should be thorough animal studies, and
3. there should be informed public support.

What do we know of the attitudes of the general public about gene therapy? Many people were surprised by a Harris poll conducted for the March of Dimes in April of 1992. It was perhaps not remarkable that 79 percent of the respondents said that they would undergo gene therapy to correct a serious or fatal genetic disease, or that 73 percent thought that potential dangers from genetic alteration of cells is so great that strict regulations are needed.

What surprised many interested observers was that 43 percent would approve of using gene treatments to "improve the physical characteristics that children would inherit," and 42 percent would support it to "improve the intelligence level that children would inherit." This high level of confidence was expressed despite the fact that 87 percent admitted to knowing little or nothing about gene therapy itself. March of Dimes officials cautioned that there is an obvious need to stress that the goal of gene therapy is "to make sick babies healthy, not normal babies perfect."

*   *   *

Those concerned over possible misuse of enhancement and eugenic manipulation can take comfort in having French Anderson as a prominent spokesperson of the gene therapy community. As the editor of the journal *Human Gene Therapy*, he has repeatedly and forcefully argued against such permutations in the genome of humans. In addition to being medically hazardous and without a justified risk, it would ask society to make moral decisions which we are not ready to make. He warns that we simply do not know enough about the complexities of the normal functions of the body to improve on what we consider to be "normal."

Anderson's position against going beyond gene therapy into any form of enhancement is straightforward. He insists that genes be used only to prevent disease and never to add what we consider desirable or useful genes to the human genome. To do so would be to compromise human dignity and open the door to increasing attempts to alter our "humanness."

Society will inevitably be asked to decide if additions to the human genome are acceptable or even desirable. We will have to decide if the risks—physical, ethical, or spiritual—are worth adding whatever gene combinations future research uncovers. The poll cited earlier indicates that many people already approve of trying to engineer our genomes to enhance our memory, intelligence, or immune system. Might it even someday be considered an obligation to use these tools to free us and our descendants from unwanted genetic limitations?

On the wall in French Anderson's meeting room when he was at NIH was a framed quotation of a few lines from *Hamlet*:

> Diseases desperate grown
> By desperate appliance are reliev'd
> Or not at all.

Surely many currently afflicted with intractable diseases desperately hope that science will find a cure for them. We have already made the transition from the recent past, when the notion of using genes as a medical tool was widely held as a "desperate appliance," to the era when gene therapy is a reality. We will soon known whether such therapy will advance to the point where the hopes of those with genetic disease will be realized.

Chapter 7

TRANSGENIC ANIMALS

The young female mouse experienced only a brief, stabbing pain as a slender, sterile needle jabbed into its abdominal cavity. A few drops of a powerful hormone, prepared from the serum of a pregnant mare, were ejected from the needle. The chemical gradually diffused into the bloodstream as the injected fluid spread slowly over the mouse's internal membranes.

The mouse was released, and it scampered off to hide under a pile of clean shavings in the cage. It was soon asleep, lulled by the soft hum of the compressor forcing a steady stream of warm air into the mouse room. The walls of the room were lined with dozens of identical cages, all bathed in the dim glow of fluorescent tubes. They delivered a calibrated amount of illumination in a constant rhythm of 12 hours of light followed by 12 hours of darkness.

Two days later the mouse felt the same piercing sensation. This time the needle delivered a human hormone, one released by the early human embryo as it first settles down into the uterus. The powerful chemicals from horse and human were foreign to the mouse but would nevertheless elicit a dramatic response. The mouse's ovaries, which before the injections had not yet been prepared to begin a reproductive cycle, were now developing at a furious pace; the female mouse had become a superovulator. As ripe eggs began to escape from her ovaries, a male mouse was placed into the cage and the two were left together overnight.

The next morning 15 to 20 newly fertilized eggs were flushed from the young female's oviducts, the tubes that lead from the ovaries to the uterus. The final blending of male and female chromosomes would not occur for a few more hours in the eggs. For now the chromosome-containing head of the sperm cell, the larger male pronucleus, lay near the pronucleus of the female egg.

The eggs were quickly placed in a sterile dish on the illuminated platform of a microscope. Peering through the lenses, a skilled technician manipulated a joystick, bringing into view a hollow glass tube whose polished tip was slightly smaller than the diameter of an egg. The tip encountered an egg, which was quickly held fast to the orifice by means of a slight suction applied to the holding tube. Another joystick drew a second glass tip into the field, a much smaller, tapered tube drawn out to a minute perforated point.

The sharp tip was plunged into the male pronucleus. A few tril-lionths of a liter of solution containing a few hundred copies of a gene was expelled, the egg was released, and another was picked up and injected. Meanwhile, a second female mouse had undergone a quite different regimen. Already sexually mature, she had been paired with a male mouse. He had been vasectomized so that, while still quite willing and eager to mate, he could not fertilize her eggs. The mating act itself triggered an automatic response in the female, adjusting her hormone levels and preparing her uterus to receive and nourish the anticipated mouse embryos.

In this case however, the embryos would not be hers, but would be the cluster of gene-injected eggs soon surgically implanted in her oviducts. Only twenty days later the surrogate mother would deliver a litter of tiny mouse pups. She would instinctively nurse them as her own. After a few weeks each would have the tip of its tail snipped off. The DNA would be extracted and analyzed. From among the microinjected eggs giving rise to the pups a few percent would have taken up the injected genes and incorporated them into their genome prior to the first cell division. As a result the new genes would be carried in all of the succeeding cells (see illustration).

Those few mice carrying the unaccustomed genetic information had become changed in a way that would have been impossible in nature's method of reproduction. Integrated into their chromosomes might now be functional copies of genes from other animals, from humans or perhaps even viruses or bacteria. The mice would be bred, and if all went well would become the founders of a whole line of descendants whose genomes had been permanently altered. They were now *transgenic* animals.

TRANSGENICS

The first report of transgenic mice derived from microinjected embryos was published in 1980. Since that time many other animals such as pigs, goats, sheep, and cows have joined the ranks of complex organisms whose natural genetic endowment has been supplemented in this fashion with genes from widely diverse organisms. Other ingenious transgenic methods besides microinjection have been developed as research and commercial laboratories across the world continue to expand the horizons of transgenics.

All of these techniques have in common a unique phenomenon—the meeting and mixing of genes from organisms which had been separated by millions or sometimes even billions of years of evolution. All of the barriers to sexual reproduction between organisms whose ancestors had long since diverged along vastly different pathways could now be breached by delivering new genes directly to a recipient's chromosomes.

Creation of new life forms is now possible, forms which could never arise by the slow, random process of mutation, sexual recombination, and natural selection. Even genes from the various kingdoms of life can be combined—plants, animals, microbes—into combinations limited only by the current crop of isolated genes, the ingenuity of researchers, and the availability of methods of predictable, efficient gene transfer.

But why do we want to do this at all? Are there any good reasons to

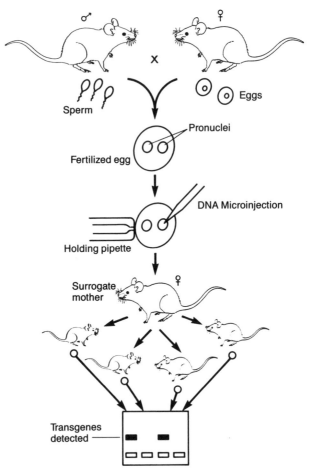

In order to make *transgenic* mice, DNA is put into the male *pronucleus* of fertilized eggs by microinjection. The injected eggs are then implanted in a surrogate mother mouse, and her offspring are checked for the presence of the new gene.

alter at such a fundamental level what appears to be the "natural" order of things? Who could benefit from such manipulations and in what way? Are there any objections to what scientists are doing as they continue to fashion these unlikely and unprecedented hybrids?

The reasons for transgenic manipulation are embedded in the fundamental aims of science—understanding and control. Human nature is marked by a burning curiosity about the world around us. Scientists are fortunate in being able to indulge this curiosity as a way of life. But modern science and medicine move beyond the familiar generalities of "scientific curiosity" and "expanding the horizons" of our knowledge to enter the sometimes problematic arena of manipulation of that knowledge for our own ends.

Welcome examples of control over nature are numerous. We have discovered that insulin regulates the entrance of sugar into our cells. With that information in hand we can control the sugar level in the blood of a person with diabetes by insulin injections. Biologists have synthesized the gaseous hormones released by female gypsy moths, and have used the tempting odors to lure the males of the species to their deaths, sparing trees from destructive defoliation. The discovery, isolation, and mass production of antibiotics, naturally occurring chemicals found to inhibit bacterial growth, has saved countless lives. Such examples are obviously the fruits of scientific discoveries.

Few would argue against the exercise of the scientific curiosity leading to research aimed at alleviating so much of the suffering to which humans have been subjected for most of our history. However, now that modern molecular science has led to intrusions which can affect the innermost control centers of living organisms and alter their very genes, such manipulations often evoke at the least a vague unease, a feeling that we are perhaps going too far.

Moreover, there are those who would argue that in manipulating living organisms to such an extent we are exercising sometimes cruel and unwarranted control over innocent creatures. Should we have unlimited dominion over the other life forms on this planet?

As usual, before we can argue the pros and cons of a particular facet of science, we need to understand the science itself. The development of the science of transgenics has been a relatively recent phenomenon, and its growth has been explosive. Annually, hundreds of reports are published describing the generation, characteristics, and utilization of transgenic animals. What are the basic principles of animal transgenics, what have been its major accomplishments, and what might its future be?

The Motives

Scientists had been unable to isolate and manipulate genes until the remarkable discoveries of the molecular biology revolution in the late 1960s and 1970s. Much of the ensuing gene-based research for most of the next decade was done with single-celled or small organisms or with laboratory cultures of cells and tissues. By the early 1980s some scientists had come to realize that the study of complex biological systems would have to move beyond these simpler models to the study of the whole organism.

Some aspects of fundamental processes such as the precise expression of genes in particular cells were being investigated by inserting genes into yeast cells or into mammalian cell cultures. But how could one get at the question of genetic control of development in higher organisms such as ourselves, where interactions among many types of cells play a crucial role? Scientists turned to their perennial favorite, the laboratory mouse. The new methods of molecular genetics were soon being used to alter the mouse genome.

Initially, the early 1980 and 1981 reports of successful DNA microinjection into mouse fertilized eggs was greeted even by many scientists as a stunt, with little significance. That perception quickly changed, however, and today most regard the work which led to those first transgenic animals as the dawn of a radically new era in biological research.

Many previously inaccessible questions about basic biological processes could now be asked, with some practical means of getting answers. What is the fate of individual cells in the early mammalian embryo? What causes the genes in various embryonic cells to function and drive the cells to develop along differing pathways? How do cells communicate with each other during development? How does this intercommunication cause some genes to "turn on," that is, to function, and others to "turn off?"

These seemingly esoteric questions are related to a quite pragmatic purpose. Transgenic animals are beginning to tell us about the molecular basis of disease. Animal models of human disease have been created— mice with cancer, diabetes, and sickle cell disease. These are being studied to ferret out the molecular mechanisms of each disease process and to test chemical agents that might control the onset of the disease, slow its progress, or diminish its symptoms. Transgenic animal models can even be used to design techniques for gene therapies directed at curing genetic disease.

Larger transgenic animals such as sheep, goats, and cows are now under development as living systems for the large-scale production of pharmaceuticals. Biologically active proteins already have been expressed in the milk of transgenic mammals. These products could be extracted from the milk and used to treat human diseases such as emphysema or hemophilia. Researchers have created transgenic pigs which can make human hemoglobin in their blood. Clinical trials to test this as a human blood substitute are planned for 1994.

Gene-altered livestock may prove to be of major importance to the agricultural industry. Domestic animals could be modified by introducing heritable genetic changes to stimulate growth or disease resistance and boost production of milk or leaner meat. Genes to improve wool growth are being sought in order to introduce them into the sheep genome. Transgenic fish have been fashioned which display a greatly increased growth rate, a crucial component in the economics of commercial fish aquaculture.

So the twin aims of understanding and control are visibly linked in the practice of animal transgenics, still only in its early stages.

The Methods

What is so special about mice? The pigs in George Orwell's *Animal Farm* established the tenet that "All animals are equal," later amended to "All animals are equal—but some are more equal than others." The "greater" equality of the mouse is based on a long history of (enforced) cooperation between mice and curious humans.

Mice are small, easy to raise and handle, and prolific. A female mouse can produce a dozen babies only three months after her own birth. Mice have a life span of only two to three years, allowing researchers to follow a developmental or disease process from fertilization to old age over a relatively short time span. Most importantly, despite the fact that our common ancestors diverged some 75 million years ago, the genes of mice and humans are remarkably similar.

Information derived from the study of mouse genes often can be applied directly to the identical genes in humans. Using our extensive knowledge of mouse physiology, we can use mouse disease models to test drug effects, devise therapies, and study the biochemistry of genetic disease in ways that would not be considered ethical or often, even possible in humans.

Extensive studies have already gone on for a number of years with mice that have spontaneously developed mutations that were passed down to succeeding generations, creating special strains of mice for research. Many of these are maintained at the world's largest mouse-breeding facility, The Jackson Laboratory in Bar Harbor, Maine, which houses over 450 mutant strains with a wide variety of genetic deficiencies. They are often aptly named. The "stargazer" mutant has a neurological defect that forces it to keep its head upturned, as if gazing at the sky. The "shiverer" mouse constantly does as its name suggests, while the "dwarf" mutants remain perpetually small because of a deficit of growth hormones.

Studying mice with these and other defects can give valuable clues as to how structure and function are related. For example, the "shiverer" mice, which eventually die of convulsions, are markedly deficient in myelin, the fatty insulating substance which covers the surfaces of nerves, particularly in the brain and spinal cord.

Others have diseases. Some have diabetes or cancer, or in the case of the widely used SCID mouse, lack the ability to manufacture T or B cells. The SCID mouse was discovered in 1980 and has become a favorite for studying the basic biology of the immune system, with particular emphasis recently on the search for control over AIDS.

There are other well-known nonrodent animal models. These are animals which have been recognized as having naturally occurring specific defects. "Watanabe" rabbits, for example, have an extremely high blood cholesterol concentration, and there are golden retrievers available for research which have muscular dystrophy. Still, scientists do not have animal models for most genetic diseases and can hardly depend on random mutation to create the kinds of animals which are now being made to order through transgenics.

A key advance in our story came in the mid-1960s. Ralph Brinster, an embryologist at the University of Pennsylvania School of Veterinary Medicine, worked out a simple method of growing fertilized mouse eggs in a laboratory culture all the way up to the *blastocyst*, a spherical 32- to 64-cell stage. Soon other researchers found that a few cells from the blastocyst of one mouse could be injected into the blastocyst of another mouse. The result was a *chimera*, a mouse whose cells are a mosaic of cells from both embryos. (These differ from transgenic animals, which have the same mixture of genes in all their cells.)

Soon, this work would give rise to the idea of using some of those early blastocyst cells as a means of introducing new genes into a mouse embryo. We will have more to say a little later about the most recent application of these so-called embryonic stem cells. The early chimeras were interesting but were not a reliable means of getting a desired genome into the germ cells, the cells which form the sperm or eggs. This meant that whatever blending of genomic characteristics might be possible with chimeras, they usually disappeared with the death of the animal.

In 1974 Rudolf Jaenisch and Beatrice Mintz, working at the Fox Chase Cancer Institute in Philadelphia, tried another approach. At that time the only sources of pure genes were viruses. They microinjected a purified monkey virus DNA into the tiny cavity in the center of mouse blastocysts. When these blastocysts were implanted into foster mothers, about 40 percent of the offspring had monkey virus DNA in some of their cells.

In 1976, Jaenisch succeeded in permanently introducing DNA into mice when he infected four- to eight-celled embryos with a leukemia virus. The DNA was permanently integrated into the germ cells and transmitted to their offspring. These were the first true transgenic mouse strains.

Finally, in a flurry of papers published in late 1980 and 1981, transgenic animal research caught the attention of the scientific community. First, Jon Gordon, Frank Ruddle, and their colleagues at Yale described the first successful DNA microinjection into male pronuclei of mouse embryos. They reasoned that since the male and female pronuclei (the nuclei of the egg and sperm) do not fuse immediately after fertilization to form the complete nucleus with its full complement of paired chromosomes, the pronuclei would be a reasonable target for introducing new genes which might become part of the chromosomes.

Then, when the pronuclei fused, the newly integrated genes would be passed on at each cell division to all the succeeding cells of the animal. The male pronucleus is larger, so it became the first target. Two of 78 mice treated tested positive for the presence of the injected DNA. The authors hailed this success as an "opportunity to study problems of gene regulation and cell differentiation in a mammalian system."

Elaborations on this theme soon appeared. Ralph Brinster and his co-workers microinjected mice pronuclei with viral DNA. This resulted in the production of functional enzymes in the mice, enzymes which had been made according to the genetic code in the injected DNA. Franklin Constantini and Elizabeth Lacy at Oxford University went a significant step further and microinjected genes for a rabbit blood protein into mouse pronuclei. Several of the resulting mice transmitted the gene to their offspring.

The stage was set for a dramatic demonstration of the power of transgenics. In late 1982 Richard Palmiter of the Howard Hughes Medical Institute announced that he and his colleagues had made "supermice." They had microinjected fertilized mouse eggs with a rat growth-hormone gene. Most of the treated mice grew two to three times faster than the untreated animals, and several ended up more than twice the normal size. The trait was passed on to their offspring.

In their December 1982 paper in *Nature* recounting the details of their stunning experiment, the scientists envisioned more than simply the possibility of developmental studies. They suggested that this technique could be adapted to "correct or mimic certain genetic diseases." Also, because exceptionally high levels of growth hormone were found in the blood of their transgenic mice, they pointed out the potential for "extending this technology to the production of other important polypeptides in farm animals." Moreover, the authors suggested the possibility of using growth-hormone gene insertions to "stimulate rapid growth of commercially valuable animals."

A year later, Palmiter's group went on to develop "supermice" using the gene not for rat, but for human growth hormone. This paved the way for the first attempt to cure a genetic disorder in transgenic animals. In 1984, Palmiter and Brinster along with Robert Hammer at Howard Hughes Medical Institute microinjected human growth hormone genes into the fertilized eggs of mouse "patients"—dwarf mice deficient in growth hormone production. The resulting mice, which would ordinarily grow to one-half normal size, now grew even slightly larger than normal.

Transgenics was firmly established as an effective technology, replete with exciting possibilities. Currently, many different genes are being routinely introduced into the eggs and embryos of mice as well as those of other animals. Most often microinjection is the method of choice. What scientific questions are being asked and answered now that we have begun to fashion organisms according to our own designs?

First, we should point out that this business of putting genes into cells is a bit more complicated than we have made it out to be. Actually, one of the troublesome limitations in transgenic manipulations is that genes do not work in isolation. They need to be in the company of other chromosomal elements, DNA sequences which may be next to the gene in question or located quite far away.

Scientists have studied gene regulation for many years in the relatively simple and easily accessible bacterial genes. As a result we really know a good deal more about what controls gene function in these primitive organisms, which do not even have a nucleus, than we do about the same phenomena in multicellular creatures, including ourselves.

We do know that genes are controlled by on–off mechanisms. A particular promoter DNA sequence is required to initiate gene function, which in turn may be regulated by other enhancer regions of the chromosome located farther away from the gene. The promoters are crucial for gene function, while enhancers are needed for maximum gene activity.

The need for this continual, dynamic and still very imperfectly understood activity of gene regulation that goes on in all living cells should be clear when we realize that each cell in an organism, with the exception of the gametes, contains all of the thousands of genes in that organism's genome. At any one time, only a few of those genes must be allowed to function, and different cells and tissues need quite different sets of active and dormant genes.

In order to add functional genes from one organism to another, the correct promoters and enhancers must either be present in the recipient organism or be added along with the new genes. As we continue to learn more about DNA sequences from analyzing the genomes of organisms, from microbes to humans, our insights into the functional elements needed for genetic transformations will take us far beyond the present possibilities.

Even with these limitations, there has been extraordinary progress in transgenics since the early 1980s. The kinds of manipulations which have been done can conveniently be divided into four areas:

1. disease models,
2. pharmaceuticals from animals,
3. animal enhancement engineering,
4. basic biological investigations.

Disease Models

In the fight against genetic disease, knowledge is the key weapon. Scientists want to find out exactly why a particular disease causes its symptoms. The answers ultimately lie in some alteration within the complex chemistry of the cells, and this in turn is based on the function, or malfunction, of genes. Effective treatment, or possibly even cure or prevention of genetic diseases depends on the intelligent application of whatever information can be gleaned from studying the disease at each of these levels.

That is why the study of animals which on rare occasions are found to have at least some of the symptoms of the disease in question, or more recently those who have been genetically altered to become symptomatic, has been considered so important in medical research. Experimental treatments can be given to these animals, and the animals can be sacrificed and examined without (for most of us) the ethical limitations attendant on working with human subjects. But animals induced to develop human diseases are, after all, still animals. They cannot be expected to be perfect replicas of humans with the same disease. For example, in 1987 two British groups independently formed new strains of mice in which the HPRT gene was no longer functional. HPRT is the enzyme defective in Lesch–Nyhan syndrome, described in Chapter 4 as a disease characterized by mental retardation and self-mutilation. However, the mice did not develop comparable symptoms.

Nonetheless, there have been a number of success stories. Let's look at a few striking examples of these new transgenic animal models.

Cystic Fibrosis. The defective gene causing cystic fibrosis was isolated and available for study in 1989. Three years later, in late 1992, elated researchers Beverly Koller and Oliver Smithies and their colleagues at the University of North Carolina announced that they had created a transgenic mouse that mimics many of the features of that devastating illness. They had succeeded in making "knockout" mice, that is, mice in which the CF gene was knocked out—literally removed. (The "knockout" technique is described later in this chapter in another context.)

The mice are most similar to humans with CF as regards their intestinal problems. They develop intestinal blockages, often the cause of death in human babies with CF. Usually these babies have mucus-clogged pancreatic gland ducts, which researchers have assumed was the cause of the blockages. However, these mice do not have severe pancreatic problems, indicating that perhaps there is more to the bowel problem than originally thought.

However, despite striking similarities in changes in the cells lining the lungs of humans and those of the CF mice, the mice do not have the mucus-clogged lungs and frequent lung infections that cause so much suffering in humans with the disease. If that can be induced, it will speed up the development and testing of new treatments for CF. For example, rather than being faced with the dilemma of having to test drugs with unknown effects on young children, perhaps on a long-term basis, large numbers of mouse models could be tested and the results could be obtained sooner, given the animals' naturally shorter life span.

Other CF models are sure to follow. The developers of this first breakthrough model are intent on making their mice readily accessible. The mice will be available from The Jackson Laboratory (Bar Harbor, Maine) for only the cost of breeding.

Sickle Cell Anemia. Five research teams have created various mouse models of this painful and often lethal genetic disease. In this condition the red blood cells, due to a defect in a single gene, are packed with abnormal hemoglobin. This condition was the first genetic disease to be understood at the molecular level, yet 30 years later there is still no effective therapy for it.

In 1990 David Greaves and his co-workers at the National Institute for Medical Research in London made transgenic mice which could synthesize human sickle-cell hemoglobin. Most of the red cells of the mice form the characteristic twisted, defective shapes characteristic of humans with the disease. Other researchers have created similar models, holding out the hope that in the near future we may be able finally to carry out the long-awaited, detailed analysis of the now-hidden factors that cause sickling of the vital red blood cells. With that knowledge in hand, there will be the possibility of developing novel, testable approaches to the treatment of this painful and often fatal disease.

Cancer. In 1971, then-President Richard Nixon declared a government-sponsored all-out war on cancer. Unfortunately at that early date the enemy was not yet in sight. We now know much more about the molecular basis of cancer. It has been revealed as a complex gene-based disease caused by mutations—frequently, by a series of mutations, some of which can be inherited.

Normal genes involved in growth and development can turn into cancer-causing oncogenes. Others which ordinarily prevent the uncontrolled growth of cells—the so-called suppressor genes—can trigger cancer if they become disabled and are no longer able to produce proteins that prevent cells from becoming cancer cells. Mutant genes often only predispose cells to malignant growth. The cancer occurs only after complex interactions with the cell's environment.

For example, breast cancer, which occurs at the alarmingly high incidence of 1 out of 9 women, is a multifactorial disease influenced by hormonal levels, diet, environmental factors, and heredity. Recent studies with transgenic mouse models have shed some new light on these complexities. Some of these mice have been made to overexpress certain oncogenes in their mammary glands, using a variety of promoters. Other mouse models display eye, heart, bone, liver, pancreas, and lymphatic-tissue cancers.

Brinster and Palmiter were the first to induce a malignancy in transgenic mice, although it was quite by accident. They had added an enhancer from a monkey virus to a gene and discovered that this addition caused brain tumors. Later, they linked monkey virus DNA to another promoter, and the cancer showed up in the pancreas.

Other studies have highlighted the effects of cooperation among oncogenes by crossbreeding transgenic mice bearing different cancer-stimulating genes. In some cases the animals that inherited both genes developed tumors at a much faster rate.

Perhaps the most famous mouse cancer model in terms of its share of newspaper headlines is the "Harvard Mouse," also widely known as the "OncoMouse." On April 12, 1988, patent 4,736,866 was issued by the U.S. Patent and Trademark Office. It was the first patent ever for an animal, a mouse developed by Harvard geneticist Philip Leder and Timothy Stewart, then at Harvard and now a senior scientist at Genentech, a biotechnology company in San Francisco.

They had been studying an oncogene called *c-myc* which was involved in certain childhood cancers. They spliced this DNA sequence into genetic material from a mouse mammary virus and used the combination to make transgenic mice which were able to pass on this genetic information to their offspring. When a female mouse of the altered strain reaches sexual maturity, the viral genes are somehow activated, and the oncogenes trigger a high incidence of breast cancer.

The new mouse has been studied intensively in an effort to track the onset and development of breast cancer and to test new drugs and therapies. While the patent was awarded to Harvard, because much of the funding for Leder and Stewart's research was supplied by DuPont, that company has commercial rights to produce and sell the animals. They are now bred and marketed through the Charles River Biotechnical Services in Cambridge, Massachusetts, where the mice are raised under stringently controlled conditions. Some are preserved as embryos stored in liquid nitrogen—in case a line should spontaneously mutate.

In 1991 Gerald Mickish reported that his research team at the National Cancer Institute in Bethesda, Maryland, had engineered transgenic mice to express a human multidrug-resistance MDR1 gene in their bone marrow. This gene codes for a protein which pumps many anticancer agents out of cancer cells before the drugs can exert their effects. Studies with these mice could lead to a way to prevent human bone marrow damage by chemotherapeutic drugs, provide a model for testing agents that could circumvent drug resistance, and offer new strategies for gene therapy.

In the final analysis, cancer can be traced to our genes. The future of cancer research will be directed at understanding and controlling those genes. The use of animal models will make an indispensable contribution to that search.

Diabetes. Our immune system is faced with what would seem to be an almost impossible challenge. Over the course of a lifetime it must recognize as foreign literally millions of different molecules and destroy them. Certain molecules, usually proteins on bacteria, viruses, and other pathogens are routinely recognized and eliminated by this complex defense. At the same time, millions of other proteins normally present in the body must be recognized as "self" and spared destruction.

At times, the immune system can malfunction, as in the case of allergies, in which there is an overproduction of the antibodies which respond to a stimulus of, for example, dust or pollen. A more serious malfunction is autoimmune disease, in which molecules of the body are mistakenly recognized as foreign invaders. Arthritis and diabetes are common examples.

In type 1 (insulin-dependent) diabetes the beta cells, the insulin producers in the pancreas, are destroyed, and the pancreas is heavily infiltrated with B cells and T cells, some of the major defender cells of the immune system. B cells make antibodies and T cells attack and destroy cellular targets. Susceptibility to this disease is influenced by a complex of genes in humans as well as experimental animal models.

Diabetes can be triggered in transgenic mice by the introduction of several different genes. In one case, an oncogene was linked to a DNA segment which promotes insulin secretion, and the gene expressed itself in the beta cells, destroying them by around five months of age. Interestingly, the diabetes formed in male mice but not in female mice.

Using NOD (nonobese diabetic) mice, Hirofumi Nishimoto's research team at Osaka University, Japan, was able to microinject a gene which prevented the onset of autoimmune insulitis, a condition similar to diabetes mellitus. Another group led by Robyn Slattery at the Royal Melbourne Hospital, Victoria, Australia, inserted a different gene and in this case protected their transgenic mice from diabetes mellitus.

Many studies continue with transgenic mice on the complex genetics underlying diabetes. These studies may not only lead to better treatment of this common and often serious disease; they may also supply scientists with mice or other animal models for testing the use of gene therapy for diabetes. There is hope that someday humans with diabetes may be given the genes they need for their own insulin production.

AIDS. Other immune system phenomena are under intense study. Certainly the most familiar of those disorders of immunity is not a genetic disease, but one in which the use of genetic knowledge could come to the rescue. It is a disease which is reaching epidemic proportions in which a virus attacks and destroys the very cells which render the immune system so effective. It is AIDS.

A major roadblock to the study of the exact mechanism of AIDS and testing for effective therapies is the lack of a suitable animal model. So far, chimpanzees and gibbon apes are the only animals that can be infected with HIV. However, even though these animals become HIV-positive, they do not exhibit AIDS symptoms.

One widely used experimental AIDS animal model, the SCID mouse mentioned earlier, is not transgenic but is a strain of natural mutants which suffer from severe combined immune deficiency. This leaves them open, as does AIDS, to any infection which happens to come along. As we noted earlier, David, the well-known "bubble boy" in Texas, succumbed to a similar condition in 1984. The SCID mice are under intensive study in the search for drugs to treat HIV and in attempts to develop an anti-AIDS vaccine.

Transgenic mice have played some part in this fight. For example, a mouse containing copies of HIV genes has been engineered whose offspring had a disease similar in many respects to AIDS. Also, transgenic mice carrying another HIV gene developed skin lesions characteristic of Kaposi's sarcoma, a type of malignancy found in about half of all people with AIDS. Scientists continue to try to create mouse models for this devastating disease.

There is a growing sophistication in transgenic animal technology. There is likewise a heavy investment in AIDS research by both the government and companies which stand to reap enormous profits from an effective treatment, vaccination, or cure for this fatal disease. With this potentially fruitful combination, there is optimism that even more ideal transgenic AIDS animals will soon be available.

And More. The list goes on. Transgenic mice have been established for, among other illnesses, certain aspects of Alzheimer's disease and multiple sclerosis, atherosclerosis (hardening of the arteries), beta-thalassemia, dwarfism, epidermolysis (an inherited skin disease), a form of hepatitis, leukemia, and neurofibromatosis, the most common single-gene disorder to affect the nervous system. By early 1993 over 100 different kinds of transgenic mice had been created in laboratories all over the world. There undoubtedly will be many more to follow.

For example, the DNX Corporation of Princeton, New Jersey, is trying to develop transgenic pigs which will supply surgeons with organs like hearts, livers, and kidneys to transplant into humans. They hope that they can engineer these pigs with genes that mask the chemical substances on cells that would ordinarily signal "pigness" should the organs be transplanted. The arrival of these foreign chemicals triggers a strong rejection response in the recipient—but only if the chemical signals of "pigness" are detected. Officials at DNX are said to be optimistic that the first swine-to-human transplants could be carried out by the late 1990s.

We pointed out earlier that transgenic animals are considered valuable subjects for testing the effects of drug treatments, trials which would be difficult or impossible using humans. Within the last few years, not only have transgenic animals emerged as pharmaceutical test organisms, but they have begun to play another role—that of producers of pharmaceuticals.

Down on the Pharm

Ever since modern biotechnology came of age in the late 1970s, when the first human protein was made by bacteria after human genes had been put into their chromosomes, a new industry has rapidly developed to the point where today, medically important substances are routinely produced by such genetically modified microbes. Genes are also put into yeast or cultured insect or mammalian cells to direct the synthesis of proteins used to treat many diseases—allergies, cancer, autoimmune diseases, heart attacks, blood disorders, infections—as well as in more mundane applications in laundry detergents or food production.

Mammalian cells have been especially useful in this regard, despite the fact that they are tricky and expensive to grow in large amounts. Because the cells are from mammals, they naturally are capable of producing the kinds of proteins that need relatively little purification or modification—a costly process usually required when the proteins are synthesized in microbes, whose cells lack the complex machinery of more highly evolved organisms. The biotechnology industry has put a great deal of effort and expense into constructing large, complex growth chambers to support such mammalian cell systems.

In 1987 the dependable transgenic mouse was once again called on for help. This time genes were inserted that caused the mice to make tissue plasminogen activator (tPA), a drug given to heart attack victims to break down blood clots that could permanently damage the heart muscle. In this case, the protein was secreted in the milk of lactating female mice. This drug was the first mammalian cell-culture product to be produced commercially; however, it would certainly take a veritable army of female mice to replace the volume of tPA which could be churned out in culture.

But what about larger animals? After all, if one could put genes into an animal which would be expressed in the milk-producing cells, why not make transgenic goats or cows rather than mice? After all, a happy dairy cow may put out more than 10,000 liters of milk annually. Being a mammal, the cow would also make proteins suitable for other mammals, such as ourselves. And of course, once a line of genetically altered cows was established, the progeny would also be able to release the product—a natural and inexpensive multiplication of living "factories."

But assuming that one could produce such livestock, would the presence of a transgenic line have any relevance to the "bottom line?" A competitive pharmaceutical company, like any other business, needs to use whatever production methods are the most cost-effective. In a few short years *pharming*—making human pharmaceuticals in transgenic farm animals—will indeed come of age. Four biotechnology companies are in on the ground floor: Pharmaceutical Proteins Limited (PPL) of Edinburgh, U.K.; GenPharm International, of Mountain View, California; Genzyme, of Cambridge, Massachusetts; and DNX, of Princeton, New Jersey. More than likely, pharmed pharmaceuticals will be on the market within five years.

PPL. Beginning in 1985, researchers have succeeded in making transgenic rabbits, sheep, pigs, goats, and cows. A dramatic breakthrough occurred with the work done by Alan Colman and his colleagues at PPL and the U.K. Agricultural and Food Research Council, both in Edinburgh. Rick Lathe, a molecular biologist, had suggested trying to apply the mouse microinjection method to sheep. He reasoned that if one could attach a human gene to one which makes a milk protein, the gene product might also be secreted in the milk.

He was correct. Transgenic sheep were produced which have either a gene for the human blood coagulation factor IX or the human gene for alpha 1-antitrypsin (AAT). Five such sheep dutifully began releasing the human proteins in their milk, albeit at low levels. Later, in 1991, the AAT level in one sheep's milk was boosted to 35 grams per liter. PPL has a multimillion-dollar contract with the German firm Bayer to bring this protein from transgenic sheep to the market as a treatment for a genetic deficiency of AAT, found in over 100,000 people worldwide, which can lead to a type of emphysema. Currently, AAT is purified from blood plasma, where it is present at concentrations of only 2 grams per liter.

GenPharm International. The transgenic pharmyard was expanded when GenPharm scientists in the Netherlands were able to modify the familiar mouse microinjection method and apply it to cows. They gathered immature eggs from the ovaries of slaughterhouse cows, matured the eggs in laboratory dishes, and fertilized them with previously frozen bull sperm. These were then microinjected with a construct of a cow milk-protein promoter and a gene for the human lactoferrin protein and nonsurgically implanted into the uterus of hormonally prepared cows.

One of the many injected eggs resulted in a male calf (which they named Herman) which showed the intact lactoferrin DNA in all his tissues. Using artificial insemination, the company hopes to produce as many as 10,000 calves per year using Herman's sperm. The females will be able to produce milk containing human lactoferrin, which GenPharm plans to add to infant formula. Lactoferrin is a protein, formerly unique to mother's milk, that slows down bacterial growth and helps a baby retain iron. Bristol-Myers Squibb of New York will have North American marketing rights to the lactoferrin-containing infant formula, which it hopes will be ready for sale in 1996.

Because the lactoferrin would be a food ingredient, the U.S. Food and Drug Administration would subject its regulation to approval through the "generally recognized as safe" (GRAS) requirements. These would necessitate that it be proven to be safe, which would require data from animal toxicity tests. Compared to the usual pharmaceutical applications, the GRAS route is shorter and much less expensive.

Genzyme. This company has formed Neozyme II, a research spin-off, to produce transgenically yet another vital human protein, CFTR. This is a molecule that regulates the chloride ion flow in cells lining the passages leading into the lungs. Defects in the genes which make this protein cause cystic fibrosis. As we have noted, this is the most common serious genetic disease in Caucasians.

However, the company has not yet made CFTR in sheep, the ultimate goal, but only in mice, using a goat gene promoter. It was released at quite high concentrations. The company plans to begin clinical trials of CFTR in 1994. It anticipates delivering the missing protein to the airway cells in an aerosol. Because the affected cells line the surface of the respiratory tract, they are accessible to such a therapeutic treatment. These cells constantly die and are replaced, so regular treatments would be needed—requiring a large supply of CFTR.

The company has also engineered goats to make tPA, the protein mentioned above which can dissolve blood clots and extend the lives of cardiac patients. Their transgenic goats, carrying the human genes for making tPA, can produce up to three grams of the substance per liter of milk.

DNX. Human hemoglobin was made in transgenic mice in 1989. In 1992, DNX scientists managed to produce functional human hemoglobin in transgenic pigs. The hemoglobin can be extracted from the pigs' red blood cells and purified. The potential market is enormous.

Each year about 70 million units of human blood are transfused, which represents about $10 billion in sales. Transgenic hemoglobin as a possible blood substitute would not carry pathogens like HIV, nor would it have to be typed before transfusion. DNX calculates that 100,000 pigs, killed and bled, would supply about $300 million of human hemoglobin. That many pigs could be bred from a single transgenic pig in five years. Alternatively, 20 pints of blood can be drawn annually from a large pig with no detrimental effects.

The hemoglobin would have to be produced at competitive prices. There is already competition from companies which make hemoglobin in genetically engineered microbes; Somatogen (Boulder, Colorado) is already testing its bacterial substitute, and Delta Biotechnology (Nottingham, U.K.) is looking at the feasibility of constructing a 100,000-liter facility for hemoglobin production from yeast. DNX plans clinical trials of its product by 1994.

In 1989 the U.S. patent for microinjection was granted to Thomas Wagner of Ohio University and Peter Hoppe of The Jackson Laboratory. They in turn assigned the patent to Ohio State University, which licensed the commercial rights to DNX. University researchers can use the technique free of charge, but biotechnology companies that want to employ it have to pay an annual license fee.

This led in part in 1990 to the awarding of a five-year contract to DNX by the National Institutes of Health to support the establishment and operation of the National Transgenic Development Facility. Researchers pay a $750 fee to DNX, which in turn will try to produce transgenic mice with the researcher's favorite gene. NIH is subsidizing the real cost of $7,000 to $10,000 per gene.

Animal Enhancement Engineering

Now that livestock are being poked and prodded into doing transgenic tricks, why not try to implant genes into these or other animals which will not convert them into walking drug makers, but will enhance those characteristics for which they are raised in the first place?

Transgenic animals have been used for the most part to study disease mechanisms or to ferret out details of other basic biological processes of life. In so doing the animals are often stimulated to differ from their normal biological performance in a negative way.

Enhancement, that is, the improvement of animals, has a goal of healthy animal subjects with good reproductive performance which will pass on any improvements to their offspring. With this in mind, transgenic rabbits, chickens, sheep, pigs, and cows, as well as fish, have already been generated.

Several major obstacles still remain in the way of perfecting this technique. Control over important traits like fat deposition, with the hope of making leaner meat, or milk production, will necessitate introducing not one gene but many. Just finding these genes is a formidable task. At this early stage, let's look at a few interesting examples of current efforts.

Big Pigs. In the United States, the gross receipts for pork add up to about $9.5 billion annually. Improvement of the rate and efficiency of body weight gain is of obvious interest to producers. Such "improved" transgenic pigs have been formed with a cow growth hormone. They show significant improvements in their daily weight gain and even have much reduced fat deposition. However, the increased level of growth hormone is detrimental in many other ways. The pigs have gastric ulcers, arthritis, enlarged hearts, and skin and kidney disease, hardly characteristics one would want to have in animals raised for human consumption. Work continues to find a system that will allow better control over growth gene expression and to find other appropriate genes for altering body fat characteristics or increasing muscle mass.

Healthy Heifers. Bacterial and viral infections of farm animals are particularly serious because the animals are often in close contact, facilitating the spread of the disease. Could we introduce genes for resistance to such pathogens? Transgenic mice bearing a newly introduced *Mx*1 gene can fight off influenza virus. Viral resistance has been engineered into a line of transgenic chickens as well. The introduced genes make only the outer protein coat of a bird virus, and when challenged with the live virus the chickens are resistant to its attack. Transgenic cows that make interferon, a protein that inhibits viral multiplication, are being tested for their ability to ward off a virus that causes diarrhea.

Scientists at Ohio University have created a virus-resistant mouse, which will be useful as a model for testing methods to create similarly resistant livestock. Despite these and other modifications, the biological processes which cooperate in the defense mechanisms against the invasions of microbes are enormously complex and incompletely understood. Because of the "cascade" effect of interdependent events, interference with one step in these processes may affect subsequent reactions in unpredictable and undesirable ways, as occurs in the transgenic pigs described earlier.

Farming Fish. When you think about the manipulations necessary to get a few fertilized eggs from mammals in order to microinject them, you can appreciate what a luxury it is to collect fish eggs for transgenic experiments. Fish obligingly squirt out their eggs by the many thousands. The eggs are large and easily collected and fertilized. Scientists are hot on the trail of domesticated fish which will grow rapidly, resist disease and be less sensitive to cold.

If you wonder why that is an alluring prospect, note that fish raised in freshwater ponds and sea pens account for 15 percent of the fish consumed worldwide, and annual sales of these top $22 billion. By the turn of the century, given the pressure on the already overworked fisheries and the growing world population, the Food and Agricultural Organization estimates that farmed fish, such as salmon, trout, and carp, will make up 20 percent of the world supply and be a vital protein source for developing nations.

Growth hormone can be fed to fish to increase their growth rate and size, but it is expensive and not necessarily efficient. In 1990 Thomas Chen and his research group at the University of Maryland's Center for Marine Biotechnology injected the genes for fish growth hormone into embryos of rainbow trout and carp. The resulting transgenic fish and their offspring grew from 20 percent to 46 percent faster than untreated fish. They became the first fish to be tested out of doors under strictly controlled conditions. Rex Dunham, a fish geneticist at Auburn University in Auburn, Alabama, is currently monitoring thousands of these transgenic creations in ponds behind the Auburn campus.

Winter flounder, which thrive in frigid Arctic waters, avoid freezing thanks to an antifreeze gene, which triggers the synthesis of proteins in the liver. These proteins suppress the formation of ice crystals in the fishes' blood. Garth Fletcher and his colleagues at the Memorial University in Newfoundland have managed to transfer this gene into Atlantic salmon, apparently enhancing the salmons' resistance to frigid waters. In 1992, Fletcher was also instrumental in accelerating the growth rate in transgenic Atlantic salmon, using chinook salmon growth hormone DNA.

Each year, the world has 90 million more people to feed. Science is beginning to look at transgenic fish as a way to help.

Basic Biology

The growing body of research related to the fundamental processes of living organisms is too diverse to attempt an adequate survey within this limited space. Actually, there is not always a real distinction between what is called basic biological research and all of the other categories of transgenic research that we have discussed above. In each case a crucial step or steps in the everyday life of an organism has been revised. The effects may be expressed, among other possibilities, as a disease or a new genetic trait; or often, in what is commonly regarded as "basic biology," a developmental path may be interrupted.

A fascinating example of the latter is the analysis of sex determination using transgenic mice. Peter Koopman from the MRC National Institute for Medical Research in London and his research group microinjected a gene called *Sry* from the Y chromosome into mouse embryos which were genotypically female. That is, the mice had two X chromosomes and were "supposed" to develop as females. The result—two of the mice were males, proving that *Sry* is the male-determining gene. This gene probably acts the same way in humans. Once the testes begin to develop, the other differences between the sexes are secondary effects due to hormones or other factors made in the testes.

Another ingenious kind of study in development involves using transgenics to selectively kill specific types of cells. Certain altered genes may be put into embryos where the new genes will function only in certain areas of those embryos. Cells in those areas are killed by toxic proteins coded for by the new genes, and the subsequent effects on development can be followed. This can be of enormous help in tracing cell lineages, that is, in trying to determine the origins of particular parts of the eye or the nervous system or any other body part.

Another approach to the study of development is to put marker genes into embryos. The cells they enter will not be adversely affected and will pass on the genes to all their descendants. Later, one can make microscope sections of tissues, stain them with a chemical which will turn blue in the presence of the marker gene, and the "family tree" of the cells is revealed. This has recently been very effective in studies of the developing retina. Similar progress has begun in marking the cell lineages in that awesomely complex organ, the brain.

The more we know about the structure and development of organisms, the better we will be able to understand how they function, from the cellular to the whole-organism level. Knowledge of structure and function is essential to the success of any attempt to control these functions, whether to alleviate or even to prevent problems which they might encounter, such as our common enemy, disease.

DESIGNER GENES

Over the past decade, microinjection of genes into the pronuclei of fertilized mouse eggs has been the method of choice which has led to a windfall of transgenic animals for basic and applied research. But despite the tremendous progress which has been made, a fundamental problem still must be overcome. When a gene is put into a mammalian cell, there is little control over where it will end up in the animal's genome. The new gene may be inserted at virtually any place in the chromosome.

In rare instances, perhaps 1 in 1,000 times, the introduced gene may insert itself into a very similar DNA sequence in the genome. This would be called *homologous recombination*. If one could somehow see to it that the cells grown into transgenic animals were those in which homologous recombination had taken place, science would have an extraordinarily important tool for studying the function of very specific genes in diseased as well as healthy animals. For example, rather than be limited just to adding genes somewhere in a genome, scientists could also knock out a specific gene by inserting a new gene in its place.

Gene targeting, that is, placing a gene at a precise spot on the chromosome could be achieved. One could knock out a specific gene, or replace it with another gene, or perhaps introduce a subtle change into a gene. Rather than just hoping to come up with a specific genetic change in a transgenic animal after microinjection, which is certainly easier than waiting for a natural mutation to create a desired change, could one hope to target specific genes as the landing site for an incoming foreign gene? Yes—gene targeting is off the drawing board and into the laboratory.

To best explain this new approach, rapidly becoming the wave of the future in transgenics, we first need to introduce a novel means of manipulating mouse embryos. In the early 1980s Martin Evans and Matthew Kaufman of Cambridge University developed a way to remove *embryonic stem cells* (ES) from 4½-day-old mouse embryos and grow the cells in laboratory culture. These still unspecialized tiny cells, growing by the millions in these cultures, retain the capacity to grow into all types of mouse cells.

When a few ES cells are removed from the culture and injected into another young embryo, the ES cells pitch in and begin to help form the developing mouse. The result is a chimera, a mouse in which many of the cells are direct descendants of the ES cells. In 1983 it was further discovered that many of these chimeric mice made gametes (sperm or eggs) derived in part from the introduced ES cells and could therefore pass the ES cell genes to future generations.

What's so important about that? Well, it was immediately apparent that using ES cells as a target for introducing new genes into embryos would offer two distinct advantages. First, rather than put genes into a relatively few cells by microinjection, other established means of getting DNA into cells might be used to expose thousands of these ES cells simultaneously.

Electroporation is now the favored method. This technique, which is also a means of inserting foreign DNA into microorganisms and plant cells, consists of exposing cells to a very brief high-voltage electric current. This temporarily renders the cell membrane permeable to large molecules. If the cells are soaking in a DNA suspension, the DNA often enters the cell. The efficiency of the new DNA actually integrating into the cellular genome is often only about 1 percent, but since many thousands of cells are being treated, many will be genetically transformed.

How does one know which ES cells are the transformed ones? In particular, how does one pick out the cells which have undergone homologous recombination?

Homologous recombination can be stimulated by using DNA which is very similar to the site on the chromosome where one wants it to integrate. Suppose that various sections of DNA on a chromosome are padlocks, and the genes we wish to insert into that chromosome are keys, each of which fits into only one of those locks. If there were only a subtle difference between a defective gene and its normal counterpart, this might allow an introduced normal gene "key" to fit the "lock" where the defective gene is located and be blocked from inserting anywhere else.

We really do not understand how homologous recombination actually works in the living cell, but it is remarkably fast, considering the size of mammalian genomes. It takes less than an hour for a DNA fragment to seek out and find its closest match among the 100,000 or so genes of the mouse.

If we assume that one begins with a gene which will often join the chromosome in a homologous fashion, the next step is to pick out the very cells in which this has happened and put only those cells into the young mouse embryo to form a chimera. One way to do this is to attach marker genes to the gene of interest. If marker genes were red lights, and one could actually see all of the thousands of cells exposed to these genes, the chromosomes of the cells where they had actually integrated would be easily distinguished from all others.

Since genes are not so easily spotted, in the real situation the marker genes usually are genes which confer resistance to certain antibiotics to whatever cells they have integrated into in a homologous way. Therefore, cells which end up after electroporation with the new DNA successfully integrated into the correct gene site can be isolated by adding those antibiotics to the growth medium. Only the cells with homologous recombination survive concentrations of the antibiotics that would ordinarily kill the cells. These cells are injected into young embryos, which are implanted into surrogate mouse mothers. Soon one obtains litters from which whole generations of transgenic mice can be raised.

There are other methods for detecting successful homologous events in the ES cells, such as amplification of the cells' DNA by PCR. Whatever the method, and others will no doubt be developed, gene targeting in ES cells is predicted to open a whole new world of possibilities for made-to-order animal models.

Exciting things are already happening. We have already mentioned the cystic fibrosis knockout mice developed at the University of North Carolina. Also, at the Cold Spring Harbor Symposium on "The Cell Surface" held in June 1992, two independent research teams reported that they had made two transgenic mouse lines to serve as a new tool for studying learning and memory. Using electroporation of ES cells as described above, they made knockout mice both lines of which are unable to make one of two specific enzymes. The mice look and act perfectly normal, but lack certain cellular responses in the brain. Those particular operations of the cells are thought to be a basis for learning, and may contribute to problems with memory. This research may be a way to get to the long-sought-after link between molecules and memory.

Over the last two years, there have been gene targeting successes with oncogenes and developmental and immune function genes as well as a variety of others. One limitation to this approach for human disease studies revolves around the fact that only a few hundred of the several thousands of human disease genes and their mouse homologues have so far been isolated. However, by 1993 more than 100 different types of knockout mice had been created in laboratories around the world, and this number is increasing rapidly.

Because mice may not adequately mimic the symptoms of all human disease, other transgenic models are being sought. ES cell lines have been established from sheep and pigs, although they have yet to be used to form chimeras.

TRANSGENIC ANIMALS: RIGHT OR WRONG?

> The question is not, can they [animals] reason? Nor, can they talk? But, can they suffer? (Jeremy Bentham, 18th century philosopher)

Public concern over the treatment and utilization of animals in experiments has had a long history. Though it encompasses many more facets of the subject than just the recent phenomenon of transgenics, the fundamental issues remain the same. They revolve around our attitudes toward creatures other than ourselves and our perspectives on where animals fit into whatever hierarchy of rights and obligations humans happen to accept.

In addition to these philosophical labyrinths, the creation of transgenic animals has introduced a unique problem: transgenics depends on the insertion of genes into animals. These foreign genes, isolated from humans or other organisms, can change the animals' characteristics in ways which would be impossible under ordinary (some would say "natural") breeding conditions, and these changes may be carried into future generations. Whatever facts, fantasies, or fears may contribute to objections to the manipulation of animals by humans is magnified enormously by the specter of genetic control. "Tinkering" with genes often conjures up a picture of scientists "playing God" by altering the very control centers of life.

Intertwined with the question of whether or not it is morally permissible to create transgenic animals is the controversial question of animal patents. In late December 1992 the U.S. Patent and Trademark Office finally ended a self-imposed moratorium and issued patents on three types of transgenic mice. These were the first since officials had approved the very first patent for a genetically engineered animal, that for the Harvard University "Oncomouse" in April 1988.

Dr. Leder, the Harvard researcher who obtained the original mouse patent, was granted another along with his colleague, William J. Muller, for a mouse that develops an enlarged prostate. GenPharm International in Mountain View, California, received a patent for its transgenic immunodeficient "TIM" mouse, of particular value for AIDS research. The third was granted to Ohio University, where researchers created a virus-resistant mouse for use in trying to develop livestock less vulnerable to disease. There is currently a backlog of almost 200 transgenic animal patent applications.

An animal patent covers animals with specific genes that do not occur in any animal species in nature and which give the animal unique, identifiable characteristics. It prohibits anyone else from using or selling these animals without permission for 17 years and may also confer rights to pharmaceutical products which the animal may produce.

Companies which create transgenic animals argue that such protection is crucially important if they are to prevent customers from, for example, buying several animals and breeding as many others as they wish. However, others see the patents as a threat to farmers. The National Farmers Union and the League of Rural Voters, for example, have warned that large corporations might end up controlling ownership of livestock, traditionally the province of the individual farmer.

In 1989 the U.S. Congress's Office of Technology Assessment

(OTA), in its publication *Patenting Life*, concluded that "It is unclear that patenting *per se* would substantially redirect the way society uses or relates to animals." The OTA has been unpersuaded by the moral arguments against transgenic animal patents and, by implication, the making of transgenic animals, for the very reason that they are moral arguments, which may make a consensus notoriously difficult or impossible to achieve.

For example, there are those who claim that animal patents would promote a materialistic conception of life and degrade humans by combining human genetic traits with those of animals. Critics also maintain that patents would also violate the integrity of animal species, that is, take away the right of a species "to exist as a separate, identifiable creature." Other arguments speak of patents as the beginning of a decline in "the belief in the sanctity and dignity of life."

The OTA report states that the moral arguments are based on "factual assertions that have yet to be proven." They likewise maintained that other kinds of arguments based on "philosophical, metaphysical, and theological considerations" were also hard to evaluate "since they usually require the assumption of certain presuppositions that may not be shared by other persons. Such arguments are not likely to be reconciled between persons holding opposing and often strongly held beliefs."

In the same publication a quite different position is presented in the statement "Consultation on Respect for Life and the Environment" signed by, among others, the National Council of Churches, The Foundation on Economic Trends, and the Center for the Respect of Life and the Environment (representing the Humane Society of the United States). Affirming that "humanity and all of nature live in a relationship of mutuality and interaction in covenant with the Creator," they urge that "a moratorium should be declared on the patenting of animals."

They object to animal patents as a "matter of deep philosophical and spiritual concern." They fear that such a practice would lead to regarding animals as "human creations, commodities . . . rather than as God's creation . . ." While they do not specifically call for a ban on transgenic animals, their statement presses for sustained public debate on such matters as "the current practice of combining human with nonhuman genetic material, . . . the probable suffering of the animals in question [and] provision for their humane care. . . ."

Meanwhile, the United States is the only country that has approved an animal patent. In fact, in December 1992 the European Parliament recommended that the European Community refrain from doing so.

These contrasting attitudes about the patenting of animals are a paradigm for the divergent approaches to the debate over transgenic animals. While there are many nuances and assumptions behind these different positions, there is a major distinction that we can make to put this into some perspective.

Concern over animal welfare is not necessarily based on considerations of animal rights. The former concern admits that doing harm to animals has moral significance, but the harm can be weighed against the potential benefits to humans, and to other animals, from such treatment. This allows at least the possibility of making transgenic animals, assuming that one takes certain precautions for the animals' care and well-being. A strict "rights" approach would posit that animals have inalienable rights which we cannot violate, and therefore creating transgenic animals is out of the question.

The latter position is rooted in a moral crusade, chronicled in detail in James Jasper and Dorothy Nelkin's 1992 book *The Animal Rights Crusade*. It grew out of a social movement which emerged in the 1970s, heavily influenced by Peter Singer in his popular 1975 book, *Animal Liberation: A New Ethics for Our Treatment of Animals*. Of central importance to Singer is that animals, just like humans, have the capacity to suffer. While not willing to grant that animals have an inviolable right to their lives, he did call for radical reform of the treatment of animals raised for our consumption.

This position did not go far enough for some. In 1983 Tom Regan, Professor of Philosophy at North Carolina State University, became a spokesperson for a fundamentalist rights position with his book *The Case for Animal Rights*. He argues that animals have an inherent worth and cannot be used as experimental subjects or resources. Even though they may lack fully human consciousness, they must all be treated as conscious beings, given the benefit of the doubt, and protected. Even if experiments such as making transgenic animals could help us in the search for disease cures or even prevention, this cannot be done "because they violate the rights of laboratory animals."

If one were to ask average citizens whether or not we should ban the use of animals in experiments which stood a good chance of enabling us to find a cure for muscular dystrophy or make a vaccine against AIDS, it seems reasonable to assume that most would support such experiments. Many would also add the proviso that care should be taken to prevent needless suffering of the animals. That position is representative of the proponents of a moderate, "welfare" philosophy.

Despite the fact that this would seem to be a mainstream attitude, many scientists are acutely aware that the more extreme "rights" position may generate enough discontent about animal research to severely curtail it. In the United States, some 400 groups opposed to animal research now have an estimated combined membership of over 10 million people. Perhaps the best known group is People for the Ethical Treatment of Animals. It claims some 280,000 members and a budget of more than $7 million and supports the dictum: "Animals are not ours to eat, wear or experiment on."

The Animal Legal Defense Fund (ALDF) based in San Rafael, California, brings the battle for animal rights into the courtroom. It has the goal of seeking "justice for all sentient beings through research, advice, negotiation and litigation." The ADLF has sued the U.S. Patent and Trademark Office to halt patenting of genetically engineered animals and is actively seeking to establish legal precedents that will help to establish laws in support of animal rights.

In marked contrast, holding the view that transgenic animals can be of great benefit to humans, the U.S. Government, many private corporations, universities, the Industrial Biotechnology Association, the American Medical Association, and certainly the majority of scientists, maintain that transgenic research should go forward.

Some, such as Jeremy Rifkin, in his 1992 book *Beyond Beef: The Rise and Fall of the Cattle Culture* and Alan Durning and Holly Brough in their chapter "Reforming the Livestock Economy" in *State of the World, 1992*, have raised more general, overarching issues. They argue forcefully that given the enormous environmental damage caused by livestock production and the fact that the nutritional energy contained in the grain now fed to animals could support many more times the number of people now fed by meat, we should question the priority given to raising animals for food.

An additional perspective is offered by Margaret Mellon, a biotechnologist with the National Wildlife Federation. She questions the validity of a high-tech "fix" for world famine. She points out that the real problem is "the lack of will to deal with complex political problems, like population stabilization, inequitable use of food resources and food distribution."

We will defer a discussion of the highly controversial issue of the use of genetically engineered animals as food until we discuss the same topic in terms of plants. Also, the wider questions of whether or not cattle-raising ought to be curtailed for reasons of environmental protection, or how we are to develop an effective political approach to world famine are thorny issues which will continue to be debated. In terms of transgenic animals, one major question is clear, although its resolution is not. Do we humans have a right to genetically alter animals for our own purposes? Should we do this to achieve the goals of understanding and controlling human and animal biological processes, including disease, to produce pharmaceuticals or other useful chemicals, and to create healthy animals with enhanced traits for our use (and profit)?

This issue must be argued with the clear understanding that in the course of transgenic animal research, although debates may rage over the level of consciousness and intensity of suffering in animals, many animals will continue to endure pain and suffering. Philosopher Bernard Rollin underlines this reality in his 1989 book *The Unheeded Cry*.

Few would question that some scientists and industries have mistreated animals. There are, of course, regulations in place, some of which have been responses to these excesses. Many carry the force of federal law (though none apply specifically to transgenic animals), which speaks to the care of animals in medical research. For example, under the federal Animal Welfare Act any institution which uses animals for research must have an institutional care and use committee. Such a panel must include, along with scientists, a concerned private citizen and a veterinarian. They must inspect laboratories at least every six months and can suspend any project if they find abuses, which they are required to report to the U.S. Department of Agriculture.

The research community also supports the efforts of the American Association for Accreditation of Laboratory Animals. This nonprofit organization encourages strict standards for the use and care of animals, "including appropriate veterinary care."

The various forms of legislation that relate to the use of animals in research are aimed at controlling pain, suffering, and distress, and where it is necessary for the aims of the research, to minimize it as much as possible. The fact remains, we inflict suffering on animals in research, some of it unavoidable if the goals of the research in question are to be met. Transgenic animals with diseases experience the pain, discomfort, or even death associated with those disorders.

The suggestion that substitutes for animals should be found does not apply realistically to transgenic animals. Other concerns such as the use of animals in cosmetics or carcinogen testing may perhaps be met by other means. However, a major reason for using animals in transgenics is that answers about the complex, integrated functions of whole organisms often requires use of the latter, and not isolated cells or tissues.

At a packed meeting of the Research Defence Society in London in April 1991, Colin Blakemore, professor of physiology at Oxford University stressed that "We must make the public realize the extent to which their lives depend on animal experiments." Nobel prize–winner John Vane maintained that "animalism" should be aligned with "quackery, cults, and terrorism." Another Nobelist, David Hubel, suggested that doctors should inform patients that the drugs which they were receiving were available only because of animal experiments.

Other speakers called on scientists to be more open about their work and to take into consideration the fact that the public is often concerned about how animals are being treated in laboratories. A year later, there was a noteworthy example of such an acknowledgment, a meeting with an ambience apparently quite different from the London meeting. NABC 4, the fourth annual open forum on agricultural biotechnology, held in May 1992 at Texas A & M University, was devoted to issues in animal biotechnology.

The National Agricultural Biotechnology Council (NABC), founded in 1988, is a nonprofit consortium of agricultural research and educational institutions aimed at providing an "open forum for persons with different interests and concerns to come together and speak, to listen, to learn and to participate in meaningful dialogue and evaluation of the potential impacts of agricultural biotechnology."

Scientists, philosophers, ethicists, and others engaged in "heated discussions" in workshops including "Links of Animal Biotechnology to Human Health" and "Biotechnology and Animal Well-Being." Immediately following the regular session, approximately fifty NABC participants—Council members, academics, business people, and others with specific interests in ethics and patenting—took part in a special seminar on "Ethics and Patenting of Transgenic Organisms."

The deliberations and divergent conclusions of that conference mirror the current status of the debates over the use and care of transgenic animals. Although many other issues were raised, NABC 4 has already been labeled as the "animal welfare" meeting. Two consensus statements of NABC 4 perhaps best summarize the scientific community's stand on the issue of making transgenic animals. They are:

1. There should be responsible, systematic investigation of the benefits and harms to animals that may be associated with biotechnology, and
2. It is acceptable under some conditions to use animals for human purposes.

Scientists have become acutely aware of people's concerns over the way science uses animals. Some scientists may dismiss these concerns as unfounded and based on ignorance. Others, more than likely the majority, acknowledge a need, as reflected in the NABC 4 "Biotechnology and Animal Well-Being" workshop statement, for developing better criteria for "responsible research and application of specific biotechnologies in animals." They add, "When there is a likelihood that a procedure will cause great suffering to animals, alternatives should be sought."

*   *   *

It seems likely that scientists will continue to make transgenic animals for basic biological and medical research and other uses. The techniques will improve. As more animal and human genes are identified, isolated, and made available, many of those genes will find their way into a variety of animal subjects.

It is equally likely that this will be done under increased scrutiny by a public wary of genetic manipulation and sensitive to animal suffering in varying degrees ranging from mild concern to rabid activism. As Bernard Rollin points out, more has been written on the search for an ideal ethic for the treatment of animals in the last ten years than in the previous three thousand. Given increased public awareness, biotechnology companies seeking venture capital and scientists seeking federal funding will probably be called on to justify their transgenic experiments with more stringent accountability.

Debates over ethics do not end in a vote. Philosophers, theologians, scientists, physicians, entrepreneurs, and others may come to different conclusions about using transgenic animals for the purposes outlined in this chapter, based on different principles and assumptions and their own interests and prejudices.

However, despite the often sincere and deeply held objections to current animal biotechnology, given the scientific, medical, and commercial gains that are now possible it would seem that there would have to be a public outcry of unprecedented proportions to turn back the clock on transgenics. Novel animals, new treatments and cures, useful products, and a wealth of information will continue to flow from the sacrifices of animals that we have chosen to be our silent servants.

Chapter 8

TRANSGENIC PLANTS

For many hundreds of millions of years the newly formed, lifeless Earth had no true soil—only air, water, and rock. Over the eons, massive geological forces thrust up mountains, which gradually wore down by weathering and erosion, while streams and rivers spread bits of their crumbling rocks across the surrounding land. Volcanoes spewed out molten rock and ash, draping a mantle of friable minerals over the waiting Earth.

Great continental glaciers crept down from the frozen North, plucking rocks from the hills and valleys and grinding them into fine clay, sand, and pebbles. Lofted by winds, fine particles of rock were spread out across vast expanses. Finally, as living organisms appeared, evolved, and spread out over the planet, true soil formation began. Plants stabilized the early soils against the wind and rain. They absorbed nutrients and water, and in a marvelous act of chemical creativity learned to drink carbon dioxide from the air and assimilate it into food, fueled by the energy of sunlight.

Dead and decaying plants became the food of fungi, bacteria, and small animals. The decomposition of plants as well as all these other creatures released the building blocks of life—nitrogen, phosphorus, potassium—enriching the soil for further growth. Their partially degraded remains lingered as humus, rich organic matter lending texture to the earth, allowing vital oxygen to pervade the soil.

For much of our existence, we humans lived as wild plant gatherers and hunters. Then, some 10 to 12,000 years ago, perhaps spurred by a growing scarcity of game due to climate changes, our ancestors began to cultivate the soil—tilling, planting seeds, and harvesting. Trial and error led from primitive techniques to agricultural practices that permitted the establishment of permanent settlements—and thus led to urban life. Now the Earth could sustain far more people than could be supported by foraging.

And so the cultivation of crops literally led to civilization. Our species now dominates the planet as no species ever has. It took from 12,000 B.C. to 1850 A.D. for the worlds' population to reach one billion. By 1930 there were two billion people. There are now 5.5 billion of us. At current growth rates the world's population will double again in another 35 years. Ninety percent will live in the Third World, already ravaged by disease, hunger, and social and political upheaval. The same extrapolation would lead to a world teeming with 694 billion humans by 2150, a little over 125 times our present population.

Regardless of the obvious conclusion that the human population must someday approach an average growth rate of zero, and leaving aside the related contentious question of what is to be considered the maximum human population that this planet can sustain, a central fact remains. World food production must be doubled if not tripled over the next several decades if the 10 billion people expected to be living by the year 2025 are to be fed adequately. Farmers will literally have to grow more food between now and then than they have since the beginnings of agriculture.

Agriculture must become more intensive and more efficient while at the same time becoming sustainable, that is, able to maintain, enhance, and not destroy the very environment that it depends on. The development of such a global system of sustainable food production is one of the greatest challenges ever faced by the human species.

This development must be brought about in a world in which increasing populations put tremendous pressures on land resources. By the year 2000 the average area of cropland per person worldwide will have decreased by half of what it was in 1950. Each year almost 3 million acres of arable land turns to desert or is lost because of waterlogging and salinity. Erosion removes 26 billion tons of topsoil annually—equivalent to the amount on Australia's wheatland. Even in the United States, after 40 years of a soil conservation program, less than one-quarter of our farmlands are under approved conservation practices.

Moreover, we face the challenge of developing a sustainable, increasingly productive agriculture in a world where rich soils are not distributed equally. One of every six humans lives in the semiarid tropics, which extend through Asia, much of India, across central Africa, and into South America. There is some limited irrigation, but farmers depend mainly on the rains, which fall for only a few weeks each year. Large-scale, high-yielding farming is out of the question, as it is for the billion people who live in other tropical and subtropical areas.

Soil and cropland loss in Third World countries is compounded by another form of environmental degradation: deforestation. As firewood is depleted, crop residues and animal waste are burned for fuel, robbing the soil of nutrients and organic matter. The soil becomes more vulnerable to runoff, soil loss, and crop damage. In the tropics the soil is shallow, its fertility dependent on constant recycling of nutrients among the vegetation. Cutting, clearing, and planting soon strips the land of these precious ingredients, and it must be abandoned.

The supply of fertile soil is finite. What remains must be recognized as our sole, precious link between the life-giving rays of the sun and the energy demands of all life on Earth. At the present levels of world food production, and assuming an equal distribution of that food, the World Hunger Program at Brown University estimates that the Earth could sustain 5.5 billion vegetarians, 3.7 billion people who get 15 percent of their calories from animal products, or 2.8 billion who get 25 percent of their calories from the latter source.

Optimists might point out that from 1960 to 1986, the amount of land on which grain was planted grew by less than 11 percent, while grain harvests more than doubled. Since 1950, rice yields in Asia have doubled and wheat production in Europe has tripled per acre, as have corn yields in this country. Food production in developing countries grew faster than population growth. However, these remarkable advances, due to improvements in crops and planting practices, have occurred in a world in which perhaps 35 million people, most of them children, die from starvation or hunger-related illnesses each year. Seven hundred million others are malnourished. India, for example, achieved food self-sufficiency in the 1970s and 1980s, and even exported food, while up to one half of its population remained undernourished, and the numbers of Indians in the lowest food-consumption groups did not decline. It is often more profitable to export crops than to sell them locally.

Hunger and starvation will not disappear solely because of increased food production. Although such increases are absolutely vital they must be accomplished in a global social context which chooses hunger reduction as a priority and not merely as one aspect of nationalistic economic and political ambitions. Agriculture must be improved in every way possible. Soil and water conservation and maintenance and enhancement of soil fertility must be aggressively pursued in a context of massive international cooperation.

A "crisis" can be defined by different criteria, but according to current estimates, approximately 35,000 children die every day because of malnutrition or diseases to which their deficient diets have made them vulnerable. It seems reasonable to state that a global crisis already exists.

This pessimistic preamble is necessary as we try to define what role our increasing knowledge and control over plant genes might play in this global drama. Recent, exciting advances have been made in our efforts to enhance the characteristics of plants far beyond their natural limitations. Our focus will be on those discoveries and what they might mean in a world expected to add 921 million people during the 90s, the largest increase ever for a decade. Ultimately, genes will not save us if we continue to multiply beyond the capacity of the planet to sustain us. But for the present, will a "Green-Gene Revolution" take over where the "Green Revolution" left off?

THE GREEN REVOLUTION

Beginning in the 1940s, the Ford and Rockefeller foundations joined forces with international governments and organizations to launch a concerted global effort against hunger that brought about what has come to be known as the Green Revolution. It was based on the application of classic plant-breeding techniques, which resulted in improved seeds, particularly of wheat and rice. These led to much greater yields per acre in regions with favorable climate, soils, and moisture if boosted by irrigation, fertilizers, and heavy use of chemical pesticides and herbicides.

Famine was indeed reduced as these intensive efforts increased the annual world grain production by a billion metric tons annually. Norman Borlaug, whose pioneering efforts had led to high-yield wheat varieties which enabled Mexico, India, and Pakistan to have record-breaking harvests in the 1960s, was awarded the Nobel Prize in 1970. However, as Ellen Messer and Peter Heywood of Brown University point out in their insightful 1988 analysis *Hunger and the Green-Gene Revolution*, this dependence on single high-yield crops and on chemicals has its serious shortcomings. There were impressive increases in the productivity of selected crop plants, but there was likewise an "unequal distribution of benefits . . . and concerns over long-term sustainability," and the changes, "by many accounts, left the poor poorer and hungrier in Asia and Mexico." They were not among those fortunate farmers with favorable land who could afford to purchase the new plant varieties and fertilizers.

During the 60s and 70s more attention was turned to less-favored regions and the need to concentrate on specific local requirements and unique characteristics of farming communities and their ecosystems. In 1971, the Consultative Group on International Agricultural Research (CGIAR) was founded. An association of 41 public and private donors, including the World Bank and the Untied Nations Food and Agricultural Organization, it supports a network of international agricultural research centers located mainly in developing countries. Over 1,600 scientists of 60 nationalities work at or for these centers, with an emphasis on developing and maintaining the long-term sustainability of agriculture which preserves the natural resource base on which farming depends.

Plant breeding has always been one of the foundations of advances in global food production. The CGIAR conducts extensive breeding programs on most of the principal food crops of the Third World—rice, wheat, maize (corn), barley, sorghum, cassava, sweet potato, and others. They also concentrate on issues which affect all agriculture: the conservation of the world's plants, both cultivated and wild, as genetic resources for future generations, the economics and politics of food production, and the strengthening of the national research capabilities of developing countries.

Recently the CGIAR focus has expanded to include additional priorities of improved productivity and sustainable production systems of livestock, fish, and forests—and to the promise of genetic engineering. At the heart of all these efforts of CGIAR, as well as those of any other agricultural research and development efforts are the domesticated animals and crops whose evolution has been deliberately altered over many centuries. Our particular emphasis here is on the genetic modifications of plants, whose selective breeding by generations of farmers has resulted in today's major food crops, which often scarcely resemble their wild ancestors.

But productivity of these plants has plateaued. Yields per acre of wheat, corn, sorghum, soybeans, and potatoes in the United States have not increased since 1970. This stasis is also true of corn, potatoes, wheat, and cassava in Latin America. Increased food production has sometimes been achieved by planting more land, not always by planting higher-yielding varieties. We must emphasize that human ingenuity in plant breeding has resulted in extraordinary improvements in crops. For example, since 1928, corn yields have increased by a factor of 5 due to the commercial breeding of hybrids (and heavy use of fertilizer).

Creating commercial hybrid seeds is a long, involved process. First, breeders seek plants with desirable traits through at least six generations of inbreeding. Some of the resulting favorable inbred lines are chosen and crossed to produce tens of thousands of different hybrids. Of these, perhaps 1 in 2,000 may be identified as a superior strain. Often, 10 to 15 years are required to come up with an improved hybrid and to prepare seed from it in commercial quantities.

Plant breeding now goes far beyond inbreeding and encompasses crossing domestic plants with wild relatives, inducing mutations in plants, or treating them chemically to make *polyploids*, plants with more the normal number of chromosomes in each cell. Such changes, while almost always harmful to animals, sometimes confer beneficial characteristics on plants.

Such efforts have been absolutely essential to the success of modern, large-scale agriculture. The cultivated varieties of plants that fed the two billion humans living in 1930 would not have been up to the task of supporting four billion just a few decades later without the several intervening decades of intensive efforts to produce plants with induced and selected genetic improvements.

Crops are vulnerable to many environmental stresses. Viral, fungal, and bacterial diseases and attacks of insect parasites are some of the natural enemies of plants that can be stymied, at least temporarily, by selecting plants that have gained a measure of genetic resistance to these threats through hybridization. Inexorably, through mutation and survival among these aggressors, nature matches the efforts of human ingenuity. New strains of organisms emerge which can penetrate the crops' defenses. For example, a new fungus strain appeared which managed to devastate much of the corn crop in this country in the early 1970s. Finally, resistant corn varieties were developed. Annually, hundreds of variants of seeds are created, sold, and planted to try to keep pace with these varying hazards of different localities.

It is not enough for a crop variety to be blessed with a set of genes which confer resistance to predators of all sorts. It must likewise produce enough foodstuff per plant to warrant its use. We are literally in a race to stay one step ahead of the natural enemies of our vital plant resources and keep pace with the Earth's burgeoning population, caught in a spiral of poverty and a deteriorating environment. In this very real scenario, plant scientists and breeders are among the unsung heroes of our society.

THE GREEN-GENE REVOLUTION

But there are natural limits to the genetic potential of plants. There are arguments over just how much more we can muster from the laborious, traditional breeding of plants. Even optimists must eventually admit that we cannot expect reshuffling of the genes of a limited number of species to somehow supply an endless stream of new and improved plant varieties.

Can we use the tools of molecular biology and biotechnology to move beyond these constraints? We have already seen how genes have begun to be recruited as weapons in the battle against human disease as well as to create unprecedented varieties of animals. Can we upstage nature in the world of plants to stave off starvation and suffering, at least in the near future? The answer is—perhaps. Extraordinary strides have been made over the last decade in the engineering of plant genomes. Just as we can insert genes from any organism into the chromosomes of an animal, we can now do the same with plants.

We are in the dawn of an era to which some refer as the "Green-Gene Revolution." Revolutions are seldom peaceful, and this one is no exception. Exciting, productive science is being accomplished in a controversial context. There are controversies over patenting and over Third World access to the fruits of gene research, as well as struggles over governmental regulation regarding the release and consumption of genetically engineered foods and products. As usual, let's look first at what scientists actually are accomplishing and then ask where it all may lead.

Plants have an extraordinary capacity for reproduction far beyond the familiar mechanisms of seeds and fruits. Back in the 1930s, plant physiologists found that root tips removed from plants would continue to grow in laboratory culture if supplied with complex nutrients such as yeast extract and vitamins. The same decade saw the discovery of plant hormones, complex chemicals produced in minute amounts by plants. Just as many of the vital functions of animals are regulated by hormones, so too are the life processes of plants.

After much trial and error, by trying various ratios of hormones in sterile culture conditions to avoid bacterial and fungal contamination, scientists found that fragments of isolated plant parts—leaves, roots, stems—would often revert to a mass of undifferentiated cells which could be grown indefinitely. Once removed from the parent plant, the cells appeared to revert to an unspecialized, embryonic condition. This *callus tissue*, growing on a solid growth medium, could also be broken into pieces and put into liquid culture where it would separate into vigorously growing single cells and clumps. All this might have been a mere curiosity were it not for the fact that this phenomenon was taken one step further—a step that has led to the possibility of creating genetically engineered plants.

The plant cells in these laboratory cultures, when given the correct combination of nutrients, hormones, vitamins, temperature, and light were discovered to be totipotent, that is, capable of growing into an entire plant. Every plant cell, just like every animal cell, has a full (diploid) set of chromosomes (with the exception, of the sex cells, which have half as many). As an organism grows from fertilized egg to adult, the genes in those chromosomes are switched on and off as the cells become specialized or differentiated into the various tissues and organs.

How these genes in cells are activated or inactivated to result in marvelously complex organisms remains one of the great mysteries of modern science. We do know that animal cells, once they become differentiated into, for example, kidney cells or lung cells, can never turn back. Even when isolated and grown in culture, they do not become other kinds of cells or tissues. However, plant cells, growing as callus tissue or in liquid suspension, can be tricked into acting as though they were fertilized eggs. Being totipotent, they are capable of dividing, differentiating—and forming whole plants.

Moreover, a callus can be cut into many pieces and each piece, given suitable growth conditions, develops into a plant. Cells transferred to the right solid or liquid medium form hundreds of tiny *embryoids*, each of which can become a mature plant.

One of the early and dramatic milestones in the discovery of this potency of plant cells was the discovery by Georges Morel in France in 1960 that the minute stem tips of orchids, carefully removed from the plant and placed in a sterile culture medium, are soon covered with dozens of tiny shoots. Each can be removed and transferred to a medium where the process can be repeated or to one in which the whole plant will grow.

These new orchids are clones, genetic copies (with some important exceptions, as we shall see in a moment) of the original plant. Once-rare orchids are now cloned commercially by the millions. This micropropagation has become especially useful in propagating plants which are free from viral infections. In India, for example, micropropagation is vital for maintaining virus-free clones of potatoes for breeding purposes.

In theory, then, any one plant can be multiplied in the laboratory into a multitude of clones of itself. As a matter of fact, tissue culture and micropropagation is still not a precise science. The exact identity and concentrations of the ingredients of tissue culture media have been worked out for many plants, but these often empirical details have been especially difficult to come by until very recently for vital agricultural crops such as rice, wheat, and maize.

In recent years, however, the key to cloning these plants, important examples of the so-called cereal crops, on which most of the world literally depends for its survival, has been discovered. It depends in part on the application of a technique developed by the English scientist Edward Cocking in 1960. He used enzymes derived from fungi to dissolve the thick cell wall around suspensions of plant cells. The resulting *protoplasts* were left with only a delicate cell membrane.

Spread out on a solid growth medium, the protoplasts first regenerated new cell walls, and then grew into callus tissue. This could then be transferred to a different food source in which new, entire plants would develop. Sometimes the protoplasts were simply put into a liquid nutrient which induced embryoids to form, from which plants could be developed.

Protoplasts can be readily prepared from many plants, but the difficulty lies in creating the suitable growth conditions for their differentiation. In 1985, rice protoplasts of the Japonica variety were the first to regenerate whole rice plants. In 1992, Indica-type rice varieties, which provide the principal food source for half of the world's peoples, were developed from protoplasts of laboratory cell suspensions.

In 1992, protoplasts from an Australian wheat, also obtained from liquid cell cultures derived from callus tissue, were induced to become normal plants. Earlier, in 1988, maize protoplasts had finally yielded whole plants. Out of the almost two billion tons of the cereal grains grown worldwide, rice accounts for 400 million, wheat for 450 million, and maize for 350 million.

With callus formation and subsequent differentiation, as well as protoplast regeneration, becoming routine in a wide variety of crop plants, the stage was set for a whole new dimension of plant improvements beyond what had traditionally been achieved by the time-honored practices of plant breeding.

New and Improved Plants

In the early days of experimentation with cloning plants from callus tissue or protoplasts it was assumed that all of the plants derived from a single source were genetically identical, just as identical twins have the same genome. It soon became obvious that there was a great deal of genetic variability among a cloned population. This *somaclonal variation* was apparently due to chromosome changes such as chromosome doublings or loss of parts of chromosomes brought on by culture conditions. These changes yielded large populations of plants which could be screened for desirable characteristics.

Such screening could be done after plants were raised to maturity, or even at the early stages of test-tube culture. Scientists spoke optimistically of selecting out drought- or salinity-tolerant plants or plants that were resistant to herbicides or various diseases. By the mid-1980s companies such as DNA Plant Technology of Cinnaminson, New Jersey, had produced new varieties of tomatoes with higher solid content, saving processors the expense of extracting extra liquid during processing to make soups and ketchups.

However, even in the most optimistic scenario, finding and propagating plants that have changed dramatically through somaclonal variation depends on random, limited, unpredictable, and undirected genetic alterations. A more dramatic means of significant genetic change was needed. Scientists turned to the protoplast.

In plant sexual reproduction, a nucleus from the pollen grain fuses with an egg nucleus deep within the ovary of the flower, and the resulting embryo is eventually released within a seed. Pollen ordinarily fertilizes a plant only of its own species, avoiding the chaos of constant hybridization in nature. In the mid-1970s, researchers found that a protoplast from any plant, no longer encumbered with a thick cell wall, could be made to fuse with a protoplast from any other plant species. They would literally blend together into one cell. Sometimes the product of this fusion would grow into a hybrid plant, a process known as somatic hybridization because somatic (nonreproductive) cells were used. The hybrids would not, however, breed and propagate their own kind.

As techniques for inducing protoplast fusion and growth became more sophisticated, it looked as though it might be possible to bypass the normal genetic isolation that ordinarily prevented gene sharing among species. Hybrids utterly impossible in nature might be created in the laboratory, mixing the traits of one plant with those of any other. Published reports of wildly improbable combinations began to appear in the literature.

Potato and tomato hybrids displaying features of both parents were formed. The more tomato-like were christened "topatoes" while others were more like "pomatoes." Even animal cells, which naturally lack cell walls, were fused with plant cells, creating, for example, living cells which were combinations of human cancer cells and tobacco protoplasts. These were interesting, even ironic novelties but of no immediate practical agricultural interest.

Once again, however, despite the fascination of overcoming the natural barriers to the mixing of genomes, the results of protoplast fusion left much to be desired. Often all or most of the chromosomes of one of the partners were lost after the fusion, and usually the fusion products did not regenerate. Still, work continues with some limited success. In fact, somatic hybridization may prove to be a valuable method for getting at least some chromosomes or fragments of chromosomes containing desirable genes from one species into the genome of another species with which it could not ordinarily breed.

Still, development of new plant varieties by regenerating callus tissue or fusion of protoplasts, while they offered many tantalizing possibilities, were soon out of the mainstream of the race to modify plants genetically. As soon as molecular biologists had learned to identify, isolate, and copy genes, the emphasis turned to the attempt to get some of those genes into plant cells. Microinjection of genes into embryos, a technique which we described in detail in Chapter 7 as the principal means of making transgenic animals, however, proved to be impractical for most plants. Nevertheless, as so often happens in science, an obscure, unheralded line of research, beginning with nothing more than the observation and study of a curious phenomenon, would offer the key to a methodology that has enormous implications.

Nature's Genetic Engineer

In 1907 Erwin Smith and C.O. Townsend of the U.S. Department of Agriculture identified the common soil bacterium *Agrobacterium tumefaciens* as the cause of crown galls, tumorous growths which often cause considerable damage to certain crops such as grapes, peaches, and many ornamental plants. Almost 60 years later, Georges Morel (of orchid cloning fame) showed that this bacterium stimulates the affected plants to make tumors that synthesize peculiar substances called opines on which the invading bacteria thrive. He proposed with great foresight that the invading bacteria acted by inserting some of their genes into the plant cells, conferring on them this new ability.

In 1974 Jeff Schell and his co-workers at the University of Ghent in Belgium proved Morel's prescience. They determined that the tumor-causing agent was, in fact, DNA. However, it was not DNA belonging to the single, large, bacterial chromosome. Instead it was DNA in the bacterial plasmids—those tiny circular DNA molecules which are frequently found in bacterial cells. *Agrobacterium* appeared somehow to be inserting these tumor-inducing plasmids into the infected plant cells.

Soon Mary-Dell Chilton and others at the University of Washington in Seattle showed that only a piece of the plasmid actually enters the plant cell. This transferred DNA (T-DNA) became integrated into the chromosome, becoming a functional part of the plant cell's genome. In other words, this bacterium, perhaps for millions of years, had been doing what molecular biologists were now hoping to do—inserting foreign genes into plants and forcing the plants to express those genes in the form of proteins.

Then, why not enlist the aid of the experienced *Agrobacterium* to ferry other genes into plant cells? By 1983 scientists had managed to "disarm" the T-DNA so that it would not cause a tumor but would still be infective. In that same year, researchers at the Max Planck Institute in

Cologne and at the Monsanto Company in St. Louis linked a gene which conferred resistance to the antibiotic kanamycin to disarmed T-DNA. *Agrobacterium* dutifully inserted this modified T-DNA into tobacco protoplasts (see illustration).

These were cultured on a kanamycin-containing growth medium. (Tobacco cells ordinarily are killed by kanamycin.) In this case, the protoplasts which had picked up the kanamycin resistance gene survived and went on to grow into mature plants. Their new trait of antibiotic resistance was of no real practical importance. But what had been accomplished was an extraordinary breakthrough. These were the first transgenic plants.

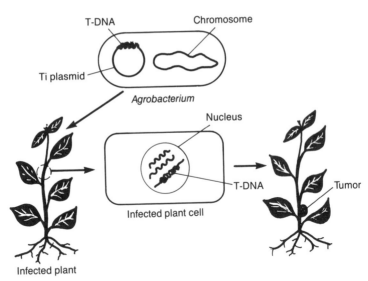

T-DNA, a small piece of the DNA of the Ti plasmid in *Agrobacterium*, enters a plant cell and becomes integrated into the infected cell's chromosomes. This causes the formation of a crown gall tumor. If the tumor-causing gene is removed, and a different gene is inserted into the T-DNA, the new gene is carried along with the T-DNA. The new gene therefore becomes part of the cell's genome, resulting in a transgenic plant.

In the decade since these landmark experiments, plasmid T-DNA has proven to be extraordinarily effective for plant genetic transformation. It is still the most widely used technique for the genetic engineering of the majority of the more than 50 species which have been genetically altered. Now the standard method for *Agrobacterium* transformation is to soak leaf segments in a suspension of the bacteria carrying the desired gene (see illustration). The segments are then transferred to a culture medium, where they form calluses which grow into new plants.

Alfalfa, carrots, cotton, lettuce, potatoes, sugar beets, sunflowers, and tomatoes are among those on the long list of new transgenic plants. However, while these are undoubtedly important agricultural commodities, they are all members of the dicot class of flowering plants. The cereal crops—rice, wheat, maize, oats, barley—are monocots. The monocots include the grass, palm, lily, and orchid families, while the more numerous dicots encompass all other flowering plant groups. Not only are the cereals not infected by *Agrobacterium*, but they also could not be regenerated at that time from protoplasts. Other methods would have to be found.

Before we look at exactly what kinds of transgenic plants have been created and what new "miracle plants" can be expected in the near future, it will be convenient to explain briefly other ingenious methods which have taken over where *Agrobacterium* leaves off.

The Direct Approach

It had already been shown a decade earlier that bacterial cells would actually pick up DNA and incorporate it into their genomes if they were simply bathed in a DNA solution. The transformation frequency was very low, but effective, considering that each drop of culture teemed with millions of microscopic bacteria. This method was modified to transform animal cells as well.

The addition of calcium and magnesium ions and polyethylene glycol (which you may recognize as the major ingredient in antifreeze) made the cell membrane momentarily permeable to DNA. Once plant protoplasts could be formed and subsequently regenerated, this direct

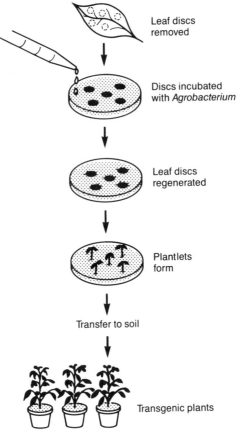

A typical procedure for making a transgenic plant with *Agrobacterium*. Leaf discs are soaked in a suspension of the bacteria carrying the new gene. The discs are then cultivated on a growth medium that allows the growth of only those plants which have become genetically modified by the procedure.

uptake technique proved to be applicable to them as well. Again, however, the number of transformed protoplasts was discouragingly low.

Another approach, also used previously for a few animal cells, again proved to be amenable to plants. It consisted in exposing a suspension of protoplasts, bathed in a DNA solution, to a brief, intense electrical current. Apparently the shock—typically 1–2 kilovolts per centimeter for 20–50 microseconds—temporarily opens up pores in the protoplast membrane and allows the DNA to enter.

This technique of *electroporation*, first demonstrated by plant scientists at the Boyce Thompson Institute for Plant Research in Ithaca, New York, in 1985, was quickly modified, refined, and applied to those plants which could be grown from protoplasts. That still left out the cereal crops. Something else was needed.

The Gene Gun

Preparing, transforming, selecting, and culturing plant protoplasts is a tedious and complicated process. However, the thick cell wall surrounding a plant cell must be removed if foreign genes are going to make their way into the cell's interior. Or is there another way? Microinjection, so effective for making transgenic animals, proved inefficient, as the fine needle tips broke or clogged as they made their way through the wall. Injecting protoplasts proved to be particularly troublesome. It was difficult to immobilize them so that they would not move during the injection, and their nuclei could not readily be seen. Many protoplasts had to be injected in order to have sufficient numbers become transformed, survive, and develop and make the lengthy procedure worthwhile.

In 1983 John Sanford, a professor at Cornell University in Ithaca, New York, had grown particularly frustrated at his inability to deliver foreign DNA into plant pollen. It seemed logical that efficient genetic transformation of pollen might prove to be a good route for passing that new DNA on to plants through subsequent fertilization by the genetically modified pollen.

He even tried using a laser beam to cut tiny holes in the pollen wall.

From those fanciful but unsuccessful attempts, however, came an idea that seemed at the time, according to Sanford, "eminently laughable." Sanford, along with Ed Wolf from Cornell's Electrical Engineering Department and machinist Nelson Allen, modified an ordinary air pistol to shoot not bullets, but a cluster of tiny tungsten particles. These particles, coated with RNA from a tobacco virus, were shot at onion epidermis. The small metallic fragments readily penetrated the cell walls, and in a few days the onion cells were teeming with the virus.

Then genes were used to coat the particles, and they, too, were blasted into the onion cells. The publication of these startling results in 1987 quickly led to the application of this particle bombardment, now popularly known as *biolistics*, not only to a host of plants, including maize, rice, wheat, and soybeans, but to animals as well. The "laughable" idea has become a quick, convenient, and powerful tool of transgenics.

In 1989, DuPont acquired the commercial rights to Cornell's gene-gun technology. Soon there were a number of modifications of this original method. In some systems, gold powder replaced tungsten and compressed helium took the place of a gunpowder charge. Stem tips, plant embryos, calluses, and callus cells in suspension have been used as targets for transformation.

Viral Vectors

Viruses, those ubiquitous infective particles that cause so much misery to humans and other animals, are potent pathogens of plants as well. A virus is not a living organism. Rather, it is merely a bit of DNA or RNA wrapped up in a shell of protein. In order for a virus to reproduce, it must first enter a living cell. If it succeeds, it may then shut down the normal chemical processes of the infected cell and turn the cell's attention to making copies of the attacking virus, which are then released, free to infect even more cells.

Virus victims are quite specific targets. Each organism is susceptible to a unique set of these lifeless parasites, most much smaller than the tiniest bacterium. Plant viruses, faced with having to attack cells encased in tough cellulose walls, make special movement proteins to ease the spread of viral particles from cell to cell. Researchers have begun to design viruses containing the genes for movement proteins in order to use the natural viral infective process to put foreign genes into plants.

At least a dozen academic and industrial groups in the United States and Europe are working on movement protein research. A dramatic proof that transgenic plants could be made by this approach was given by scientists at Biosource Genetics Corporation of Vacaville, California, and the University of California, Riverside. Using genetically modified viruses, they induced tobacco plants to make the protein tricosanthin, an experimental AIDS drug. Just as animals are now being used to make pharmaceuticals, transgenic plants may soon be manufacturing these as well, as we shall describe shortly.

DESIGNER PLANTS

Having looked at the principles of some of the current methods for making transgenic plants, let's turn to a summary of their application to date, with particular emphasis on vital crop plants. Just as for other aspects of genetic engineering, as we have pointed out, we have only begun to tap the potential for changing nature to meet our needs (and desires, not necessarily synonymous). The creation of the first plants of the genetic future has been met with great fanfare on the part of many, particularly those who will profit by their use, and with fervent opposition by others.

If one could design the ideal crop plant, it would grow rapidly into a robust specimen, laden with nutrient-rich seeds and fruits, despite having to grow under stressful conditions such as periodic drought, salinity, mineral deficiencies or toxicities, or temperature extremes. It could get its critical nitrogen supplies by extracting nitrogen from the air, rather than depending on fertilizers. As it grew, using energy from an optimized photosynthetic system, it would rebuff the attacks of insect, bacterial, fungal, and viral predators.

This is the plant described in a variety of scientific publications in the early 1980s, spurred by the newfound possibilities of plant genetic transformation and the growing facility in gene manipulations. A major barrier then and now to the realization of this ideal has been our limited understanding of the molecular bases for the expression of plant genes, let alone the fact that the genes for most important plant characteristics have not yet been isolated.

Looked at from the perspective of the hopes generated by the rapid development of plant genetic transformation techniques, the fact that we do not as yet have an abundance of transgenic plants that can, for example, pull nitrogen from the air for fertilizer, require less water, or have markedly improved nutritional qualities is disappointing. On practical grounds, however, a proliferation of plants with those kinds of genetic alterations will not be available until the genes controlling these characteristics are identified, isolated, and made available for research.

The search for these genes is under way, but in the meanwhile, a large number of transgenic crop plants have become a reality, although they are not as yet commercially available. The novel traits genetically engineered into them are those for which the genes have been identified. The majority are crops that have been altered to withstand their perennial enemies—weeds, predators, and disease.

It is not only the scientific complexity of plant transgenics that has led to the current emphasis on the genetic engineering of resistant plants. We already have many excellent, useful crop plants that are the products of generations of classic plant breeding. However, plant diseases alone—the debilitating attacks of bacteria, fungi, and viruses—lay waste, on average, 12 percent of crops worldwide. It makes sense to try to maximize conditions for the plants we already have.

Also, the future availability of the transgenic plants described below requires a testing, evaluation, and approval process that can last from four to ten years or more. Economic considerations dictate that companies in the *agbiotech* (agricultural biotechnology) industry must identify a profitable market and invest in development of a product with a reasonable chance of financial return. The current state of progress in agbiotech is a reflection of those needs. There is certainly a market for resistant crops, and the genetic technology for producing such plants is in place, and expanding.

Herbicide Resistance

"Weeds," remarked nineteenth-century naturalist Ralph Waldo Emerson, "are plants whose virtues have not yet been discovered." Since Emerson's time, however, agricultural scientists have waged war against weeds, hidden virtues notwithstanding. Herbicides, or "weed killers," comprise the bulk of all pesticides used in the multibillion-dollar agrichemical industry.

Weeds compete for moisture, nutrients, and sunlight, and can reduce a field's potential yield by up to 70 percent. Tilling to uproot weeds leads to soil erosion. Faced with losing vital soil, large-scale farming has turned to increased dependence on herbicides to control weed growth. More than 100 types of herbicides are now used amid increasing concern over their toxicity to animals, including humans, and their persistence in the environment.

Herbicides are plant killers whose heavy application may ironically damage the crop as well as the weeds. There have been two general approaches to get around this dilemma. One is to alter the level or sensitivity of the herbicide's specific chemical target within the plant. The other is to put into the plant genome a gene for an enzyme that inactivates the herbicide.

Scientists at Monsanto and at Calgene in Davis, California, have made transgenic tomato, soybean, cotton, and other crops that are resistant to glyphosate, the active ingredient in Roundup, a Monsanto product. This is currently the most widely used weed killer because it is active in low doses and is rapidly degraded by soil microorganisms. Roundup kills plants by inhibiting the action of the enzyme ESP synthase. The transgenic plants have been given genes which make large amounts of a similar but much less sensitive enzyme, which literally uses up the herbicide in its futile attempt to damage the crop.

Calgene also has modified cotton with a bacteria-derived gene which allows the plants to be sprayed with bromoxynil, an herbicide produced by Rhone–Poulenc. Calgene officials expect that their herbicide-tolerant cotton will be approved for sale in 1994.

Likewise, researchers at DuPont have used a similar approach to engineer canola and cotton that can tolerate sulfonylurea compounds, the active ingredients in the firm's herbicides Glean and Oust. At Plant Genetic Systems in Ghent, Belgium, and at the German company Hoechst, the second approach to herbicide tolerance was successful. Scientists isolated a gene for an enzyme from a bacterium that inactivates glufosinate, the active ingredient in the herbicide Basta. Transgenic glufosinate-tolerant canola, soybeans, and corn have proven themselves in field trials.

Disease Resistance

A wide variety of bacteria, fungi, and viruses are natural enemies of plants. Significant progress has been made against plant viruses, a persistent problem in crop production, particularly in the tropics, where viral infections may wreak extensive damage.

A standard genetic trick, termed *cross-protection*, is to infect plants with a mild strain of a virus which protects the plants against damage by more potent strains. In 1986 Roger N. Beachy and his co-workers at Washington State University in collaboration with Stephen Rogers and Robert T. Fraley at Monsanto introduced a gene which triggers the production of the outer protein coat of the tobacco mosaic virus (TMV) in tobacco and tomato plants. They become strongly resistant to TMV infection.

Since then, scientists have stimulated effective tolerance to more than a dozen viruses in a number of crop species including alfalfa, potatoes, melons, and rice. There have been some successes against other pathogens as well. Researchers in Japan have placed a gene into tobacco that renders it resistant to the so-called "wildfire" disease, caused by a common soil bacterium. The gene apparently allows the plant to make an enzyme which destroys a toxic substance produced by the invading bacterium.

In July 1990 a patent was issued to DNA Plant Technology Corporation of Cinnaminson, New Jersey, for a method of genetically modifying plants with bacterial genes that make the enzyme chitinase. This enzyme breaks down chitin, a major component of fungal cell walls. The method is designed to protect crops from fungal attack, particularly during the long trip from farm to market.

Insect Resistance

Farmers annually lose billions of dollars worth of crops because of insect predation. Their major weapon against these invaders has been chemical pesticides. A strategy to reduce dependence on pesticides by creating insect-resistant plants has come from a study of the bacterium *Bacillus thuringiensis* (Bt). It makes an insecticidal protein that binds to receptors in the gut of caterpillar larvae, disrupting their ability to feed. Extensive toxicological tests over 30 years have led many to label Bt-based products as the world's safest insecticide.

Applications of dried preparations of Bt have been used for many years as a temporary protection against caterpillar infestations. In the mid-1980s genetic engineers at Monsanto, Plant Genetic Systems, and other companies isolated from the bacterium the genes responsible for making the insecticidal protein. They soon had put the gene into tomato, potato, and cotton plants. Use of these plants is predicted to reduce the use of insecticides on cotton by at least 50 percent.

Other forms of Bt genes have been found which extend its range beyond caterpillars. Transgenic Russet Burbank potatoes have been made which are effective against the Colorado potato beetle. Recently, scientists at Mycogen Corporation in San Diego have isolated Bt genes that are active against plant parasitic worms, and others have identified Bt genes active against mosquitoes.

Resistant Plants of the Future

As bizarre as it may sound, plants develop increased immunity to disease when treated with aspirin. In 1990 it was discovered that plants under attack use a biochemical alarm signal that triggers a variety of defense mechanisms throughout the plant. The messenger, produced by the plant in response to attacking disease organisms, is salicylic acid, a close relative of aspirin. Researchers at Ciba–Geigy laboratories in Basel, Switzerland, have identified a set of genes that become active when plants are sprayed with salicylic acid. They have been dubbed "SAR genes" because they seem to stimulate a broad range of defense responses in the plants, conferring a systemic acquired resistance (SAR). The precise function of the genes is not yet known.

Virtually all plants respond in this way when sprayed with the acid or aspirin. When attacked by animal pests, the responses are different, diverse, and complex. They include increasing the levels of toxic substances in the leaves as well as production of another signal chemical, called systemin, which hinders the ability of insects to digest the plant.

Simply spraying plants with systemin or aspirin is not enough to protect them. These substances are quickly metabolized by the plant. But knowledge of the genetics of this fascinating system of self-defense may lead to a means of immunizing plants against their enemies, just as we stimulate the human immune system to protect us from pathogens that cause polio, tetanus, or diphtheria.

Food Processing

In addition to the early successes in protecting plants against their natural predators, a few notable victories have emerged. While it is the lot of fruits to rot, any delay in their ripening and eventual senescence would be welcome. In one crop alone—tomatoes—30 to 40 percent are bruised during harvesting, shipping, and storage and consequently spoil. The search is on for the genes that are turned on or off by bruising.

We know that bruising triggers production of the gas ethylene in the fruit. Ethylene is also produced normally during ripening. The shelf life of tomatoes can now be extended for several weeks by two genetic modifications of this phenomenon. Athanasios Theologis and his colleagues at the Plant Gene Expression Center in Albany, California, operated by the USDA's Agricultural Research Service and the University of California, have blocked the expression of a gene in a metabolic pathway leading to the synthesis of ethylene.

They used *antisense* gene technology (see illustration). This works

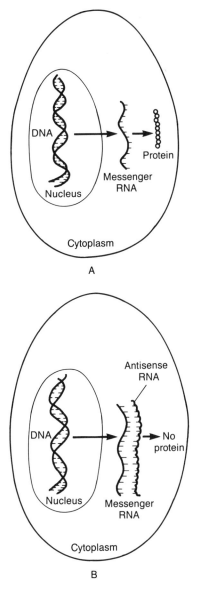

A

B

In this form of *antisense* genetic engineering, the messenger RNA is blocked from using its information to synthesize a protein by binding it to RNA with complementary bases.

by interfering with the way information gets from the genes to the protein-manufacturing sites in the cell's cytoplasm. Messenger RNA, the molecule carrying the genetic message from the DNA, can be blocked if a gene making DNA or RNA that will bind to the messenger RNA is put into the plant's genome.

Another version of antisense methodology has allowed Calgene, Inc. of Davis, California, to engineer a tomato, dubbed the "Flavr Savr," which stays fresh for two weeks after vine-ripe harvesting. This transgenic tomato was modified with a gene which expressed antisense RNA, blocking the synthesis of a softening enzyme which breaks down pectin in the cell walls of ripening fruit. Calgene expects to market the new tomato in the late summer of 1993.

Improved Nutrition

The transgenic plants discussed above are those which are probably closest to marketing approval. Meanwhile, reports continue to be published on a wide variety of plant characteristics. For example, genes have been isolated which code for proteins rich in methionine, one of the amino acids (building blocks of proteins) that humans cannot synthesize, but must include in our diet. Scientists at Monsanto Corporation have expressed a bacterial gene in potatoes giving the tubers a 60 percent increase in starch content. If they can likewise raise the starch levels in corn or other grains, they could reduce the cost of ethanol, a fuel additive made from these plants. It could also lower the cost of biodegradable plastics made from starch.

The major seed companies are devoting significant resources to the development of genetically modified crops. Pioneer Hi-Bred, the nation's leading seed vendor, is working to create modified grains and oilseed crops, such as low-saturated-fat canola, soybeans, and sunflowers with enhanced oil and protein qualities. They are also aiming for alfalfa and corn with altered protein and starch content. In addition, Pioneer is modifying the nutritional characteristics of feed crops so that farmers can feed their animals without having to add protein supplements. For example, they are developing soybeans for poultry feed with higher amounts of key amino acids that normally have to be added to the diet.

DeKalb, the second-largest seed-producer, is working on producing feed corn with more nutritious proteins as well as higher oil content. Other gene manipulations at DeKalb are aimed at fashioning corn ears producing improved edible and industrial oils, and strains with more starch for industrial cornstarch users.

We have only begun to realize the potential for creating more nutritious and hardy plants for foods and useful plant products.

Manufacturing Plants

In 1989 a "third wave," as Robert Fraley of Monsanto has expressed it, of plant biotechnology got under way. Plants are now being engineered to produce, not only foodstuffs, but specialty chemicals.

Calgene, for example, is engineering the canola genome for more than food. The firm intends to prod canola to contain high levels of erusic acid, an industrial oil used in the manufacture of rubber and plastics and as a cutting lubricant for steel and metal work.

If transgenic animals can be made to produce pharmaceuticals and other useful, commercially valuable proteins, why not try to modify plants to do the same? After all, although many of these same products can be made in genetically engineered bacteria, fermentation vats of microbes might churn out gram quantities. Fields of transgenic plants could manufacture hundreds of pounds of extracted proteins per acre.

Plants also have an advantage over animals in that their "scaleup" time is faster. That is, plants can be grown and harvested in a few months compared to sometimes several years before transgenic animals produce the foreign proteins in their milk.

The first, surprising demonstrations that this might be possible came in 1988, when scientists at the Friederich Miescher Institute in Basel, Switzerland, introduced the gene for human interferon into turnip plants. This protein is part of the body's natural defense mechanism against viruses. The turnips proceeded to make large quantities of human interferon.

At about the same time, Andrew Hiatt and his co-workers at the Research Institute of Scripps Institute in La Jolla, California, fashioned transgenic tobacco that dutifully produced antibodies—which the researchers dubbed "plantibodies." The tobacco and more recently alfalfa plants are under development as possible commercial sources of antibodies for diagnostic and therapeutic uses, or possibly as supplements for infant formulas.

Antibodies are proteins which bind and inactivate specific antigens, usually other proteins. Because these antibodies can be made to stay within the plants' cells, the binding ability of the antibodies within the cells of plants might be employed to recover contaminants or valuable compounds from the environment.

It has become obvious that plants, like other organisms, can make human proteins, given the proper genes. In 1989 researchers at Plant Genetic Systems and at the Rijks University, both in Ghent, genetically engineered the oilseed plant and *Arabidopsis*, a plant which has become favored for many genetic studies, to produce seeds which contained leuenkephalin, a human brain chemical. Another group at Mogen International in Leiden, the Netherlands, induced the growth of human serum albumin in potatoes and tobacco. Solutions of this protein from other sources are used widely to replace fluid in burn patients and others.

In 1992 Agricultural Genetics Co., Ltd., in Cambridge, U.K., revealed that it had devised a way to produce large quantities of animal vaccines in plants. After cowpeas (known as black-eyed peas in the U.S.) were infected with a genetically engineered cowpea virus, the plants made proteins from animal pathogens, including the serious foot-and-mouth disease virus. One leaf from an infected plant produced enough protein to make 200 doses of vaccine.

Currently, bioengineered vaccines are made in bacteria or cultured animal cells. In contrast to the tedious procedures accompanying this approach, plants are much easier to grow, the isolation of their proteins is simple, and the yield is high. The company's next targets are to induce plants to assist in making vaccines against hepatitis, warts, and the common cold.

Other intriguing results include tobacco plant cells from Biosource Genetics of Vacaville, California, which make melanin when growing in liquid culture. The melanin, a black animal pigment usually extracted from octopi and squid, completely blocks harmful ultraviolet radiation. It can be used instead of PABA, now the sunscreen of choice.

The above results underline the enormous potential of plants as factories for making proteins to order. However, more than likely the first marketable chemicals from transgenic plants will not be pharmaceuticals, which need to undergo lengthy testing for safety and efficacy. Perhaps they will be specialty chemicals for the industrial, commercial, or food markets. For example, consider the development heralded by front-page headlines in the April 24–26, 1992, issue of *USA Today*: "Researchers sow seeds, harvest plastic."

The headlines were referring to the work done at Michigan State University and at James Madison University in Virginia. *Arabidopsis* was genetically modified with two bacterial genes responsible for the bacterium's curious ability to make polyhydroxybutyrate (PHB). This is a biodegradable plastic, a polyester used for making plastic bottles at a present cost of about $12 a pound. The plants made the plastic in a granular form. Plant manufacturing of this plastic, if perfected, is expected to cost about $1.20 per pound, which will encourage more widespread use of this environmentally friendly packaging material.

Oleochemicals are oil-based chemicals with a wide variety of uses in foods, cosmetics, soaps, lubricants, and other products. Currently, oleochemicals are derived from petroleum or plants such as tropical palms. Now, the most important oilseed crops grown in Europe and North America–soybean, rapeseed, and sunflower–are being genetically modified to produce new types of these oils. For example, Calgene, Inc., is testing a transformed variety of rapeseed which is high in stearic acid, allowing the oil to be low in saturated fats. This could be used for margarine, chocolate, and other confections. Others of their rapeseed oils could substitute for the coconut oils in shampoos and detergents.

Other kinds of food products, at least in terms of food additives, are on the horizon. Calgene has engineered the production of cyclodextrins, novel carbohydrates with potentially high value for flavor and odor enhancement, and the removal of undesirable compounds (such as caffeine) from foods. They added to potatoes a bacterial gene that responded by converting some of their stored starch to cyclodextrins.

In May 1992 Lawrence Berkeley Laboratory and the University of California at Berkeley, California, researchers reported that they had fashioned transgenic lettuce and tomato plants that can make monellin.

This is a protein that has a flavor 100,000 times sweeter than sugar, which might make it possible to alter the sweetness of edible plant products without raising the sugar content.

Or perhaps the first commercial success will be unusual flowers. DNA Plant Technology Corporation was awarded a patent in July 1991 on a genetically engineered white petunia, in which the gene producing a purple pigment was turned off. And in August of that same year, after a four-year search, researchers at Calgene Pacific in Melbourne, Australia, isolated the gene that controls blue coloration in petunias and irises. They plan on using the gene to create transgenic blue roses. Cut flower sales are a $5 billion worldwide market, of which roses account for about $500 million.

*   *   *

And so we are in the formative stages of what is already a revolution in the useful and lucrative manipulation of plant characteristics. The optimism of the early 1980s envisioned the genetic transformation of our major crops into more drought- and stress-resistant, highly nutritious plants, with enhanced photosynthetic capabilities and the capacity to extract ("fix") nitrogen from the air to literally help provide their own fertilizer. These miraculous plants would not only enhance the agriculture of the industrial world, but also transform farming in less-developed nations.

The reality of the early 1990s presents a different scenario. True, the more complex traits such as photosynthesis, drought resistance, and nitrogen fixation are each controlled by a complex of genes and are not so readily manipulated as the mainly single-gene characteristics described in this chapter. And as we have seen before in the case of the development of transgenic animals, the complex tasks of locating, isolating, and cloning genes as well as gaining an understanding of the ways in which they can be turned on or off in cells is likewise an essential prerequisite to gaining control over specific plant functions.

But has the recent dramatic progress in controlling the plant genome at least been aimed at achieving the ideals of a sustainable world agriculture—one which maintains and enhances the environment? Has it focused on creating plants to provide nutritious food sources, plants readily available to farmers in underdeveloped nations?

In an editorial in the June 1992 issue of *Bio/Technology*, editor Douglas McCormick pointed out that wheat had finally joined soybeans, rice, and corn among the major crops which have been genetically transformed. He reminded his readers that agriculture is the world's largest industry, but did not "boast the pharmaceutical business' profit margins or high technology—yet." He went on to ask, "Does anyone want to be left behind if the titan stirs? . . . with the conquest of wheat, the most important principalities of a vast kingdom lie at our feet."

Is this vision of conquering nature compatible with a nurturing stewardship of the environment and the hope of more abundant food for the hungry poor? There are those for whom a "conquest" metaphor conjures up images of plunder and booty, hardly compatible with a sustainable agriculture. For others, the profit motive is seen as necessary if costly research and development is to be carried out on any plant genetic modification.

Such research and development, whatever the motive, has engendered other controversies. Since 1989, there have been over 500 applications pending with or approved by the United States Food and Drug Administration and the Environmental Protection Agency for field testing of genetically engineered organisms. Most of these have been plants with enhanced disease, herbicide, and pest resistance. Critics argue that once these genetically novel organisms have been moved from the laboratory to the field, whatever risks they may pose to the environment can no longer be controlled. Unlike a car model with a faulty transmission, they could not be recalled.

Could transgenic plants transmit their genes to other species by cross-pollination? For example, would a cross between a crop and nearby wild weedy relatives result in more pernicious weeds? Or could some of them escape the farm and become weeds themselves? Could the widespread use of pest-killing plants select out insects naturally immune to the plants' defenses and neutralize the intended benefits? What about the safety of food prepared from transgenic plants? Could a gene alteration interfere with a poorly understood metabolic pathway and cause the plant to produce a substance that might prove dangerous to consumers?

Some would dismiss these concerns as fanciful, given what they consider to be adequate regulatory safeguards against such dangers. The public perception of plant genetic engineering may be quite different however. For example, in July 1992 more than 1,000 American chefs pledged that they "would not serve genetically altered foods in their restaurants." According to one prominent New York chef, "I owe it to my customers to serve them pure, nutritious food of the best quality possible."

In the case of genetically engineered organisms, can we distinguish between fact and paranoia? How is their testing and release actually regulated? Will a safe and affordable supply of plants from a "green-gene" revolution soon be available to help stave off malnutrition and starvation?

Chapter 9

FIELD OF GENES

"We are all ignorant," a philosopher once wrote, "but from different perspectives." There are certainly widely differing perspectives from which to view the promise and perils of transgenic organisms. Such gene-altered creatures are in the problematic arena of biotechnology.

While biotechnology has its share of unalloyed enthusiasts and vocal opponents, it is probably safe to say that scientists as a whole might best be described as "cautious optimists" toward its role in our future. This is, of course, a generality, and one could conveniently cite uncritical enthusiasts and apocalyptic naysayers from among the ranks of working scientists.

Some would argue that biotechnology has arrived just in time to enable us to make a quantum leap up and away from many of the biological constraints and environmental hazards we now confront. According to these projections, transgenic crops offer the promise of high-yielding, disease- and pest-resistant, nutritious plants raised with a minimum of chemicals and posing no threat to people or to the environment.

Others argue that the creation and field testing of such transgenic plants may cause unacceptable risks. Moreover, critics maintain that the subsequent patenting of these plants will lead not to a "kinder and gentler" agricultural sector, but to one in which the giants of the agricultural-chemical and seed industry stand to reap enormous profits. They point to the current emphasis on the development of herbicide- and pesticide-resistant plants. These critics maintain that such efforts are directed not at reducing the use of these substances, but at increasing the market for crops that tolerate even heavier use of the chemicals produced by the same company that furnishes the seeds.

Based on our experience of the "bottom-line" approach of much of American industry during the Reagan–Bush era, which led to increasing environmental degradation, there is precedent for such pessimism. On the other hand, in their public statements many industry representatives are adamant about their intent to balance their intentions to develop and aggressively market transgenic plants with a sincere regard for safeguarding human health and the health of the environment.

In the background lies the widespread public perception of gene-altered crops as inherently suspicious entities. The Food and Drug Administration made a ruling on May 26, 1992, which it claimed reflected the consensus of the scientific community. It declared that genetically engineered food is not intrinsically dangerous and would not need special approval before entering the market. Moreover, the FDA would not require labeling of genetically altered foods.

Katherine Matthews, staff attorney for the Pure Food Campaign, responded that the FDA policy is "garbage. It requires nothing. It's appalling. We want testing before the food is out there. We want it labeled. And then we want you to boycott it."

The Pure Food Campaign was launched by environmental activist Jeremy Rifkin and others in direct response to the FDA announcement. They demanded that the FDA undertake mandatory premarket testing of all genetically engineered foods, label them, and require premarket public notification by all manufacturers of the presence of such foods in stores, so that the foods could be traced if illnesses or other problems arose.

The prominent New York restaurant chefs who joined the controversial Rifkin at a press conference in June 1992 affirmed their support for what the organizers hoped would be an international boycott, when and if genetically engineered food enters the marketplace. As one chef put it, "I will not sacrifice the entire history of culinary art to revitalize the biotechnology industry."

Where does the truth lie? Or rather, is there a truth that we can discern while trying to distinguish between fact and fantasy? The only way to find out is to look at the record of actions as well as words, and listen to the arguments. It is difficult, however, to be a completely impartial observer of the struggle to balance profit making with risk taking. The abundance, variety, quality, and safety of the world's food supply affects each of us and all our descendants.

Biotechnology is driven by economics. So far, because of the availability of the necessary genes, as well as market demands, most agricultural biotechnology research and development has centered around the genetic modification of plants to create varieties that are resistant to herbicides, insects, and plant diseases, have a longer shelf life, and reduce processing costs. Such plants will be valuable commodities. Given the profit motive, it is perfectly logical that the first transgenic plant headed for market is the Flavr Savr tomato and not a high-protein crop for the world's impoverished hungry. Calgene, Inc., invested $20 million and eight years in this plant, which they expect will capture 15 percent of a $3.5-billion market.

This is simply a model of private enterprise at work. It should be no surprise to anyone who has ever bought a new car model or the latest computer. However, among the serious concerns raised about this situation is the notion that the economic incentives to create, develop, test, and market genetically engineered organisms have created a climate in which the health of the environment as well as individuals could be seriously threatened, perhaps in ways that may be irreversible.

Let's look at three prominent current issues in agricultural biotechnology which illustrate these concerns. The first involves plants which have been engineered to make their own pesticides. The others center around the controversies over the release of transgenic plants into the environment, and their use as food. All three issues have yet to be resolved.

A PEST KILLER: A PARADIGM?

In recent years, many farmers have begun to see their dependence on chemical insecticides in a new light. There has been a shift towards biological control agents—naturally occurring defenses against these major plant predators. Many are convinced that there are environmental and health risks associated with repeated exposure to some pesticides. Moreover, in the United States, in spite of a tenfold increase in the use of chemical insecticides from 1945 to 1988, crop destruction by insects rose from 7 to 13 percent. Over 500 species of insects have developed resistance to one or more chemical insecticides.

This resistance has developed from the very same mechanism that drives biological evolution. When a large population of pests is first exposed to a new chemical poison, most (or sometimes all) are killed. By chance there may be a few which have a natural resistance to the chemical, rendering them less sensitive to its effects. For example, they may be able to make relatively large quantities of enzymes which can detoxify the agent, and thus survive.

Repeated applications result in classical Darwinian natural selection. The susceptible weaker insects die and the stronger, resistant ones remain to pass on their survival genes to the next generation. Eventually, the resistant forms predominate in the population and the insecticide becomes useless.

For many years, a naturally occurring insect toxin has been a popular biological control agent for destroying larval forms of insect pests in agriculture and for tropical disease control. Fifty years ago, *Bacillus thuringiensis* (Bt) (see Chapter 8) was just another of the many bacteria found in soil—until it was discovered that it makes toxic proteins that, when eaten by certain insect larvae, destroy their guts. But it is harmless to humans. By 1938 farmers had begun spraying dried Bt over their crops as a safe, minimally disruptive pesticide. It often became a substitute for other chemicals to which pests had become resistant.

But Bt has several drawbacks. Various forms of Bt are active against only a narrow group of insects and has to be applied often to remain on the plants at effective concentrations. As a result, Bt has held less than 1% of the agricultural insecticide market. The biotechnology industry in recent years has responded by developing new sprays that last longer and contain higher levels of the toxin, spawning optimistic projections that Bt sales could grow to top $1 billion by the turn of the century.

Moreover, even after years of use, there did not appear to be any Bt resistance developing among insect populations. It was assumed that the subtle but rich diversity in Bt toxin types along with their natural (bacterial) origins rendered Bt immune to insect resistance.

The promise of Bt soared to new heights in the late 1980s, as genetic engineers isolated the bacterial genes responsible for making the toxic proteins and transferred these genes into plant genomes. Tobacco, cotton, tomato, and potato varieties were now armed with Bt genes, which would render the cells of those plants capable of manufacturing the bacterial proteins and obviate the need to spray the fields with the bacteria themselves. Even more insect types could be killed, since now not only leaf-feeding insects but sucking, sap-feeding, and root-dwelling larvae would be targeted.

By 1992 one-fifth of all the applications to field-test transgenic plants in this country involved crops which had been given Bt genes. Monsanto announced plans to market its Bt-containing transgenic cotton seeds by 1995.

But even before these genetically modified plants had become a reality, there had been disturbing reports of the appearance of Bt-resistant insects. Beginning in 1983, first on the U.S. mainland and later in Hawaii, Thailand, the Philippines, and other sites, farmers and agricultural scientists documented examples of an unprecedented loss of Bt effectiveness. Ironically, in using massive doses of Bt, environmentally conscientious farmers had unwittingly set the stage for natural selection. In 1989,

Monsanto scientists showed that the tobacco budworm, a major cotton pest, had developed resistance to Bt when exposed to heavy doses of Bt in the lab.

Thus, one of first transgenic plants headed for market was in trouble. It was exposed as a sheep in wolf's clothing. While initially effective against insect predation, the genetically engineered plants ran the risk of becoming ineffective as their very presence favored the survival of resistant insects. As resistance spread, we would begin to lose a valuable genetic resource—insects that are susceptible to Bt. According to *Genetic Engineering News*, a leading biotechnology newsletter, "the industry . . . must take swift steps to save its products."

Citing predictions that widespread planting of Bt-producing crops would render Bt useless as a biological control agent and drive farmers to return to chemical insecticides, some critics called for a halt to the planned release of transgenic Bt crops. Faced with losing a potentially enormous market, the biotechnology industry responded by taking immediate steps to assure that Bt remains effective in the near future—by planning a technological "fix."

For example, Monsanto plans to advise future users of their Bt-producing cotton to use crop rotation and carefully planned planting and plowing schedules to minimize insect exposure to Bt. The firm also may sell mixed seed lines, that is, Bt-producing and non-Bt-producing seeds within the same bag of seed. That would create "refuges" of non-Bt plants in a field. The susceptible insects living on those plants could breed with their resistant fellows, and many of the next generation would inherit the susceptibility genes.

Meanwhile, hopes of a transgenic solution remain within the industry. Increasing knowledge of how the toxin actually works at the molecular level may lead to genetic engineering of even more potent varieties of Bt for use after insects gain resistance to the currently used compounds. Also, work is under way to alter the expression of the Bt genes in transgenic plants. Perhaps they might be maneuvered into being active only in certain tissues or during critical stages of plant growth and not during the entire growing season. Still other genes might be added with different modes of action, making it unlikely that insects would develop resistance to several gene products simultaneously.

How will all this play out? Certainly the companies that envision a major market for transgenic Bt seeds are not about to halt their efforts. Of course, it would be counterproductive for them simply to sell seeds that would soon become ineffective. In January of 1992 the United States Department of Agriculture sponsored a meeting of government and academic researchers who fashioned a complex strategy which they hoped would save Bt as an effective insecticide.

One of their principal recommendations was that there be an independent national advisory board to serve as a Bt guardian. It would report to the USDA and recommend avenues of research on the bio-pesticide and monitor its use. However, at a "Bt Resistance Workshop" sponsored a few months earlier by the National Audubon Society, Ann Lindsay, director of the Environmental Protection Agency's registration division in the Office of Pesticide Programs, told disappointed workshop participants that the EPA had no plans to regulate the use of new biopesticides.

No enforceable federal regulatory program is in place to ensure that proposed management strategies are adopted by growers and seed companies. The National Wildlife Foundation has recommended that neither the EPA nor the FDA approve large-scale tests of Bt plants until the federal government has developed a plan for delaying by 20 to 30 years the development of Bt resistance in insect populations. Until that happens it appears that farmers will have the ultimate responsibility for resistance management. They will decide whether and how to use Bt products.

Meanwhile, there are reports of a new approach. Genetic engineers are aiming at inserting insecticide-resistance genes into insects. Beneficial insects such as bees that pollinate crops or ladybug beetles that eat other insect pests are often killed by the commonly used broad-spectrum chemical insecticides. Populations of transgenic chemical-resistant bees and beetles could be released while farmers continued to apply the chemical insecticides. Farmers could have their (chemical) cake and spray it, too.

Some would say that the Bt story is a paradigm for the pitfalls facing the promises of genetic technology. Others look at it as a technical challenge arising from the inevitable struggles of a technology still in its infancy. Whatever the case, it does raise the question of the role of regulatory oversight of an industry preparing products with far-reaching effects.

WHO'S MINDING THE STORE?

Fortunately, the environmental legislation passed by Congress in the 1970s, together with other federal laws, provides ample protection of public health and environmental quality. . . .(Industrial Biotechnology Association, *Agriculture and the New Biology*, 1987)

This four-billion-dollar industry should grow to fifty billion by the end of the decade—if we let it. The United States leads the world in biotechnology, and I intend to keep it that way. (then-President George Bush, 1992)

Simply stated, this society—its profit-oriented market, powerful corporate and military interests, and orientation toward dominating nature—cannot be trusted with biotechnology. (*Biotechnology: an Activist's Handbook*, 1991)

As with all contentious issues, one's attitude is influenced by, as the saying goes, where one is "coming from." That is not to imply that all opinions are necessarily flawed or insincere. It does suggest that one proceed beyond a sometimes emotionally satisfying argument to an analysis of the facts. It may turn out that careful scrutiny of the facts ultimately may reveal a position, viewed by some as extreme, as the correct one.

The overarching issue here is the need to ensure the safety of the environment—the air, soil, and water of the Earth as well as its human, plant, animal, and microbial life—without denying this planet and its inhabitants the full benefits that could accrue from biotechnological advances.

In the case of the controversies over governmental regulation of biotechnology, in particular the engineering, field testing, and use of transgenic organisms, there are clearly divergent opinions from which to choose. The data needed to support these opinions, however, are sometimes difficult to come by, because they frequently involve conflicting interpretations of the very same data. A major stumbling block to consensus is that many of the issues under the umbrella of environmental stewardship are based on predictions. Will there be any serious, perhaps irreversible harm if transgenic organisms are moved from the laboratory to the field? Can these genetically engineered organisms be eaten by humans or domestic animals without unacceptable risk?

Could engineered genes be transferred into other organisms, transforming them into pernicious pests? For example, could a crop resistant to herbicides and producing pesticides pollinate a related wild, weedy species which might prove difficult to eradicate?

Ideally, answers to most of these and many other related questions would depend heavily on having recourse to a system of *predictive ecology*. This would be a well-developed science which could project with a high degree of certainty how a living organism will behave, where it will survive and multiply, and if it does, what effects it will have on other living organisms, including humans. Such certitude is not now possible, and it would take an enormous investment of time and money to amass and analyze enough data to satisfy the most ardent critics that the risks in any given case were minimal.

Some critics would argue that, because science does not have such a system, extreme caution should be used against the possibly irreversible effects of releasing organisms from the laboratory into the wild. Once set free, they, or other organisms to which they might transfer their engineered genes, would become a permanent part of the world ecosystem. Others, predictably among them the producers and potential sellers of genetically engineered organisms, support caution but stress that perfect predictions can never be possible.

They are willing to cooperate with regulatory demands for careful testing of their modified organisms, but maintain that to wait until we have a complete and infallible means of prediction, even if that were to become possible, would be to stifle the incentives to create organisms which might now play a vital, beneficial role in alleviating the scourges of hunger or disease.

What has been the attitude of regulatory agencies in this country towards these questions? What role have they played in defending the

integrity of the natural environment and our health and safety while
trying to maintain international leadership in commercial biotechnology?

Field Testing: "Planned Introduction"—or "Release"?

It seems more realistic to think in terms of fairly small chances
(perhaps roughly calculable) of less-than-catastrophic problems that
one could overcome or live with. The risks in prospect are likely to be
no different in kind from those that have always been associated with
the introduction of the novel products of plant and animal breeding.
. . . (*Genetically Engineered Organisms: Benefits and Risks*, University of Toronto Press, 1991)

People will pay for this hundreds of thousands of years from now.
. . . Every introduction is a hit-or-miss ecological roulette. (Jeremy
Rifkin, 1985)

A new era opened in American agriculture on October 9, 1985,
when a 27-year-old veterinarian, Roger Saline, methodically injected a
live, genetically altered virus into 250 piglets on a farm in Bradford,
Illinois. The USDA had given the go-ahead for genetically engineered
life forms to be deliberately set free from the laboratory. Public disclosure
was not made until April of 1986.

On May 31, 1986, scientists from Agracetus, a small biotechnology
company in Middleton, Wisconsin, planted 200 four-inch transgenic
tobacco seedlings on a front-yard-sized plot somewhere in south-central
Wisconsin. The company would not disclose the precise location of the
site because it feared the experiment would be destroyed by protesters.
The outdoor planting, the first ever of a gene-altered plant, had been
approved by the Environmental Protection Agency on November 13,
1985.

One day later, the EPA likewise approved the first deliberate release
of genetically engineered microorganisms into the environment. They
authorized Advanced Genetic Sciences (AGS) of Oakland, California, to
conduct field tests of bacteria designed to prevent frost damage to
strawberry plants.

Two months later, officials from AGS made an embarrassing revelation. Their antifrost-bacterial experiment, which had been temporarily prohibited by a local Monterey, California, ordinance because of growing concern over unknown hazards to public safety, actually had been secretly carried out over a six-month period the previous year—before the firm had received EPA approval. The company had injected the bacteria into trees at its headquarters, but had "followed all the guidelines . . . to the letter." John Bedbrook, science director at AGS, admitted to a House Science and Technology subcommittee that the company had shown "a real lack of wisdom" although it had acted "in good faith."

All of this was going on as federal administrators awaited new policies from President Reagan about which government agencies would oversee genetically engineered products to be used by farmers and how the agencies would do so. Arguments over the safety of the release of gene-altered organisms from the confines of the laboratory to open fields continue to the present. Meanwhile, under a still-evolving set of federal and local regulations, permission has been granted to conduct more than 900 such tests.

The fundamental document outlining the roles, responsibilities, and policies of the Federal agencies involved in biotechnology was published in June 1986. The Domestic Policy Council of the White House released the *Coordinated Framework for the Regulation of Biotechnology*. According to David T. Kingsbury of the National Science Foundation, the purpose of the document was to explain to the public that "for questions involving the products of biotechnology (more specifically, organisms derived from recombinant DNA technology), human health and the health of the environment were of paramount concern and were adequately protected."

Some of the document's premises have guided subsequent policy. It stated that existing laws were for the most part regarded as adequate for biotechnology oversight and that most biotechnologically altered organisms are not fundamentally different from nonmodified organisms.

Most biotechnology products intended for release into the environment are or will be regulated by legislation that was already in effect before the arrival of modern biotechnology. These laws were originally intended to protect the environment from plant pests, chemical contamination, and pathogens. Implementing these statutes is among responsibilities traditionally assigned to several agencies, such as the EPA and the USDA.

Under the *Coordinated Framework*, various agencies were given regulatory authority over particular categories of transgenic organisms. The USDA has the responsibility for biotechnology in agriculture and forestry. Within the USDA, the Animal and Plant Health Inspection Service (APHIS) is designated the lead agency for regulating plant and animal biotechnology products. APHIS conducts an environmental assessment for each proposed release. If APHIS judges that there is no significant risk to the environment or to agricultural crops, a permit is issued. The decision to grant a permit, which frees an applicant from having to make a full-scale environmental impact statement, is based on already-established regulations under the Federal Plant Pest Act of 1957 and the Plant Quarantine Act of 1912. Both of these statutes were intended to prevent the entry or dissemination of living organisms considered dangerous to American agriculture.

The overall philosophy of USDA is expressed in the National Academy of Sciences (NAS) 1987 publication *Introduction of Recombinant DNA–Engineered Organisms into the Environment: Key Issues* and in the National Research Council (NRC) 1989 document *Field Testing Genetically Modified Organisms: Framework for Decisionmaking*. They maintain that the product and not the process should be regulated. In their judgment, the products of biotechnology do not differ in important ways from unmodified organisms or conventional products.

Regarding the risk of testing transgenic organisms in the environment, in the words of Robert H. Burris, chairperson of the NAS panel that created the 1987 statement, "We feel fairly confident that if this thing is done right, it will not pose any hazard. . . . We hope that this will be reassuring to the public."

In 1988 the NRC had published a "white paper" claiming that field tests of novel organisms are for the most part safe. It met with objections even from *The New York Times*, which criticized the paper for the absence of evidence to support its reassurances. The 1989 report provided detailed documentation as well as a framework of questions designed to help regulators assess the risk of proposed experiments.

This NRC publication recommended an approach that had been proposed earlier in that same year by the highly respected scientific organization, the Ecological Society of America and had been welcomed by environmental groups. The NRC proposed weighing on a case-by-case basis the characteristics of each organism and its potential effects on the environment, with no reference to the means used to modify the organism genetically.

A 1988 OTA report, *Field Testing Engineered Organisms: Genetic and Ecological Issues*, reiterated the attitude of this congressional advisory board toward environmental release:

> Although there are enough uncertainties that introductions should be approached with caution, a large body of reassuring data, derived chiefly from agriculture, supports the conclusion that with the appropriate regulatory oversight, the field tests and introductions planned or probable in the near future are not likely to result in serious ecological problems.

Earlier, in 1985, the White House had convened an advisory body, the Biotechnology Science Coordinating Committee (BSCC) to "monitor the changing scene of biotechnology and serve as a means of identifying potential gaps in regulation in a timely fashion. . . ." The BSCC was a forum for senior policy officials to identify and respond to key issues. However, in the polite words of the Office of Technology Assessment, this committee "had difficulties in obtaining consensus on important issues such as risk assessment [and] levels of oversight. . . ."

The White House dismantled the BSCC in 1990 and handed over the original charge of the BSCC to the President's Council on Competitiveness (COC), chaired by then–Vice President Dan Quayle. The Council's stated purpose was to improve the plight of American business in the global market. Critics were quick to suggest that this, to put it mildly, was not a deliberative body predisposed to caution when confronted with the pressure to expand our markets for biotechnological products.

A report by OMB Watch and Congress Watch, two groups associated with well-known activist Ralph Nader, had pointed out at least a dozen areas where the COC had stymied federal regulations. It had, according to these critics, dramatically reduced the protection of the nation's wetlands and gutted key provisions of the Clean Air Act. Although George Bush had campaigned as the "Environmental President," his attitudes toward environmental protection soon took on a deregulatory bent reminiscent of the Reagan years.

The COC did not disappoint its critics. In 1990 and 1991 the Council proposed a series of new recommendations designed to block regulations that would "discourage or penalize innovation." According to these policies, genetically engineered organisms should "not be subject to federal oversight" unless there is substantial evidence that they present "unreasonable" risks.

Even among government agencies, there was dissension about the COC proposals. EPA officials were reported to have been stunned when they saw the proposed guidelines. They had been in the process of updating their own regulations, which would at least have notified the government of new transgenic products, which could then be analyzed individually for possible risks.

The biotechnology industry met the proposals with mixed reactions. The COC had recommended that the various federal agencies implement the criteria for oversight in their own ways as they categorized organisms according to the risks associated with environmental release. Industry officials preferred clearly defined regulatory criteria, enabling companies to estimate accurately the time and costs of seeking approval for testing and marketing their biotechnology products.

Then, on February 1, 1991, USDA published a nine-page document, *Proposed Guidelines for Research Involving the Planned Introduction into the Environment of Organisms with Deliberately Modified Hereditary Traits.* Its principal intent was to assist in the design of safe field trials. The National Wildlife Federation assailed the guidelines as inadequate, a policy of "let's wing it and hope for the best," rather than a rigorous monitoring and record-keeping system. It pointed out that the document had no enforcement provisions, giving the investigator complete discretion to decide where a particular organism fits into the regulations. The USDA, of course, defended its guidelines, while adding that the White House had insisted that the guidelines be published separately from any plan for implementing them.

In his January 1992 State of the Union address, President Bush announced a moratorium on new environmental health and safety regulations, including those covering biotechnology. Then, on February 27, 1992, the COC issued a report that listed new biotechnology guidelines. In the words of the COC's press statement, it was a plan for "streamlining federal regulations."

It placed the burden on the regulator to show evidence of a hazard rather than on the producer to show evidence of safety. It also advised regulatory agencies against applying stricter regulations to the altered organisms than apply to the parent organisms when the new product does not have new traits known to be more dangerous than those of the organisms from which they are derived. The COC document had no strict legal authority, but skeptics pointed out that the White House has arbitrary veto power over all proposed regulations through the office of Management and Budget.

In the words of Daniel Grossman, an editor of *geneWatch*, a newsletter published by the Council for Responsible Genetics (CRG), "the assumption of *safe until proven otherwise* has left a regrettable legacy of hazardous chemical products and waste." He added that the document "presumes, mistakenly, that agencies know how to evaluate whether or not a new trait increases the risk of a novel organism." The CRG is a Cambridge, Massachusetts, based national organization of scientists, public health advocates, environmental activists and others who want to see biotechnology developed safely and in the public interest.

On the other hand, a March 1, 1992, *New York Times* editorial hailed the COC for its "scientifically sound approach to the tricky business of deciding which products are worth strict regulation and which are not." The writer did admit that whether or not the COC would head off efforts to crack down on unsafe products would be unclear "until agencies try to issue new regulations conforming with the policy."

*   *   *

Staunch critics of loosely regulated oversight of field testing prefer the term *release* to denote moving genetically modified organisms from the lab to the field. The term tends to conjure up provocative images of letting loose some perhaps uncontrollable and potentially harmful wild creature. Defenders of relaxed government standards often prefer the term *planned introduction*. This suggests that a field test is analogous to a blind date—filled with potential for a bright future while admitting the possibility that it might just not work out.

Whatever we call open-air trials of transgenic organisms, should they be feared? What are seen as the risks of putting transgenic organisms out into nature, now that they have functional genes from other, usually unrelated organisms, creating genetic combinations impossible by natural means of reproduction?

As usual, the seriousness of the risks is seen in the light of one's perspective. Some argue that when new technologies are proposed, the burden of proof that they will not cause unacceptable harm should fall on the proposers. Many biotechnology enthusiasts argue that fears of widespread harm are hypothetical and deal with the possible, not the probable. They point out that there are no examples of environmental problems associated with the several hundred transgenic organisms field-tested since 1984. They prefer to be presumed innocent, leaving the burden of proof on those who create scenarios of disaster.

Wherever the obligation of proof lies, the real difficulty comes in determining how to arrive at that proof in a way that is both scientifically valid and capable of engendering public confidence in the conclusions. In its August 1992 report to Congress, *A New Technological Era for American Agriculture*, the Office of Technology Assessment offered its opinions on two categories of risks which might be posed by the release of large numbers of genetically engineered organisms into the environment, with particular emphasis on plants. The following is a summary of this statement.

> *Crops Becoming Pests.* Certainly there have been instances when a species accidentally or deliberately introduced into a new environment has become an undesirable, uncontrollable pest. Dutch elm disease, which has destroyed most of our native elms, is a consequence of the accidental introduction of a fungus. Strawberry guava, introduced into Hawaii as a crop, escaped from cultivation and is now a serious weed pest in natural habitats. Kudzu vine, which now runs rampant in the

South after being brought in as a ground cover, is another deliberate introduction gone awry.

Based on many years of farming experience, it is highly unlikely that an agricultural crop would escape the fields to become a trouble-some weed. DNA-modified plants are designed to thrive only in the rich and nurturing environment of a cultivated field. Many traits for hardi-ness outside the farm have been bred out of crops. Studies have shown that serious weeds have an average of 10 to 11 "weedy characteristics." Crop plants have an average of only five such traits.

However, until we know more about the consequences of large-scale use of transgenic plants, we need to use deliberate caution rather than be complacent.

Gene Transfer. Many of the newly conferred genetic traits of transgenic crops—herbicide-, disease-, or pest-resistance—would be very beneficial to weeds as well. If cultivated crops have an opportunity to cross with related wild, weedy relatives, the engineered genes could be transferred to them. Should that risk deter us from creating and testing such crops?

True, the possibility of plants cross-breeding exists. However, the danger can be eliminated or reduced to acceptable levels by several means. These include keeping the transgenic plants far enough away from their wild relatives that their pollen cannot reach the wild plants or harvesting the crops before they flower, thus eliminating pollen produc-tion. Also, the crops could be further engineered to have reduced pollen production or be given genes that would reduce the ability of the pollen to fertilize other plants. Another strategy would depend on engineering new characteristics such as herbicide tolerance into plants that would require two genes on separate chromosomes to be effective. This would dramatically increase the odds against the transfer of this trait to another plant in the wild.

In any case, most crop species in the United States are grown far from the plant relatives from which they were derived, although particular care would have to be taken in the case of canola. This important oilseed crop does have the potential for hybridizing with closely related species. The developing countries are the center of origin for many of our crop species. Precautions must be taken by the federal government, companies, foundations, and international agen-cies to offer risk management advice for these situations.

Well-designed, carefully monitored field tests of increasing size and complexity should allow negative impacts to be detected while there is still an opportunity to correct them.

Critics point to what they consider several weaknesses in the above arguments. Some take more seriously the history of foreign organisms invading our ecosystems. They point out that 11 of the most serious weeds of the world are crops in other regions. In the United States, of a total of 5,800 introduced crops, 128 species of agricultural and ornamental plants have escaped from the cultivated fields and have become pest weeds.

In April 1992 experts from the Center for Environmental Management at Tufts University, Medford, Massachusetts, released their careful scrutiny of a representative sample of environmental assessments issued by the USDA in granting permits to field-test transgenic plants. In 32 percent of the USDA assessments there was no discussion of potential weediness, even though all of them stated that the risk of transformed plants expressing such plant-pest characteristics would be examined.

APHIS takes the position that if the plant is not considered a weed before it is genetically transformed, then changing one of its genes cannot transform it into a weed. The Tufts analysts point out that engineered characteristics such as tolerance to insects or disease would seem to give the new plant a competitive advantage.

On the question of the possibility that foreign genes in modified plants might be transferred to plants outside the test site, they warn that the distances chosen to isolate these experimental sites from other plants might not be great enough, based on recent studies. And while most of the assessments considered the likelihood of hybridization between the crop plant and wild relatives, there were several omissions. For example, while two assessments examined proposed field tests of transgenic cotton in Hawaii, they did not include any specific information on the potential for cultivated cotton to pollinate the wild native cottons known to inhabit the islands.

Further, they point out that, despite the fact that the APHIS assessments described methods to dispose of transgenic plants and to detect and destroy any sprouted seeds that remain in the soil, they do not require information on the cold-hardiness of the seeds or the length of time they can survive in the field. Nor did APHIS address the potential movement out of the test site through the activity of animals such as birds, rodents, or ants.

Certainly, the flow of genes by pollination from crops to natural populations of plants is far from well understood. In March 1992 Norman Ellstrand, an ecologist and professor of genetics at the University of California at Riverside, published his studies of the spread of genes from a plot of cultivated radishes to wild radishes planted up to a distance of one kilometer away. He found that the gene flow even into the most distant plants was much higher than predicted, leading him to suggest that isolation of field tests from compatible weeds would require at least several kilometers.

The OTA report cited above candidly admits that in the matter of forming regulations for environmental protection, "the balancing of risks and benefits is difficult and open to bias." The challenge, as it points out, is to determine those products to be used or activities planned which "would have adverse effects significant enough to warrant legislation."

In that determination we see again a tension between the need for careful oversight of biotechnology and the pressure to encourage successful application of that technology to create products that will enhance U.S. economic competitiveness. Given the complexities of trying to predict all possible ecological harm from field tests with our current limited state of knowledge and the fact that so far as we know no adverse effects have yet occurred, any concerted attempt to call a halt to field tests would probably be unsuccessful.

By 1993 over 900 applications were pending or had been approved by the USDA or EPA for releasing genetically engineered organisms into the environment—almost all of them plants (see illustration). By now presumably those and other releases have been conducted, since the USDA and EPA generally approve pending tests. Unlike automobiles with defective transmissions, transgenic plants that could cause environmental harm could not necessarily be recalled. The USDA continues to permit an increasing number of field tests with growing confidence. Meanwhile, the regulatory framework for judging those applications is still evolving.

A modification in the APHIS guidelines again came under attack by environmentalist critics. In October 1992 the USDA was assailed for approving a petition by Calgene, Inc., which exempts the company's genetically engineered tomatoes from regulation under the Federal Plant Pest Act. The USDA decided that Calgene had shown that the plants did

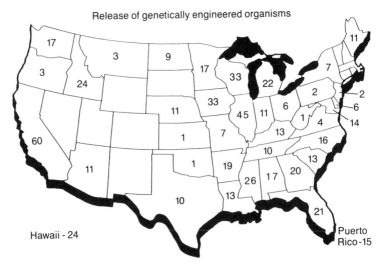

Release of genetically engineered organisms

Hawaii - 24

Puerto Rico - 15

This map shows the number of applications pending or approved by the USDA or the EPA for testing genetically engineered organisms as of October 1, 1992. The figure may underestimate the number of tests in each state, as applicants may test in more than one site in each state per application. By May 1993, the number of total tests had risen to 966. We assume that approved tests have been conducted. The USDA and EPA generally approve pending tests. (From *The Gene Exchange*, October 1992, p. 12. Reprinted by permission of the National Wildlife Federation. ©1992 National Wildlife Federation.)

not pose a risk to plant or animal health. This would allow the company to sell the Flavr Savr tomato anywhere in the United States.

Registering its disapproval, the National Wildlife Federation pointed out that APHIS really did not have an adequate database from which to judge the safety of the approved field tests. A 1992 NWF survey had shown that only half of the scientists who had conducted field tests on a wide variety of plants from 1987 to 1990 had at the time of the survey submitted the required final reports to APHIS. The NWF also objected strongly to a petition to the USDA by the Upjohn Company to deregulate the company's genetically engineered virus-resistant squash. The NWF pointed out that Upjohn had performed no field tests to address ecological concerns. In this case, the USDA refused to deregulate the transgenic squash plants.

Also in October 1992, the Department of Agriculture began a 60-day public review period for another proposal to relax certain regulations. By late 1992 a number of university scientists were clamoring for a relaxation of the APHIS case-by-case assessment of field trial applications. A September 4, 1992, editorial in *Science* argued that experience had shown that "crops thoughtfully modified by rDNA [i.e., transgenic plants] do not become or create new plant pests."

On March 31, 1993, the USDA issued rules replacing field test permits with a notification procedure in the case of six types of genetically modified crop plants: corn, potatoes, tomatoes, soybeans, cotton, and tobacco. Under the new regulations, researchers working with these plants will simply have to notify APHIS of their intent to transport or field test the plants and certify that they will abide by published procedures and performance standards.

There are several qualifications. For example, the process must not create a disease in the plant, and the new genes put into the plants may not be functionally intact animal or animal disease—causing genes. According to APHIS, 85 percent of field tests that currently require extensive review before permits are given could be conducted under the notification procedure. They note that this should reduce the cost of the preparation of a permit from $5,000 to $500. Biotechnology officials were quick to hail the prospect of a new system of streamlined regulations.

Of course, enforcement of even the most stringent federal regulations stops at our borders. The global environment observes no such boundaries. How are the concerns over release of genetically engineered organisms being addressed outside of the United States?

Worldwide, there have been three basic approaches to regulation. A number of countries with active investment in biotechnology have no specific biotechnology regulations. This includes most of the growth-oriented countries of the Pacific Rim, such as Taiwan, South Korea, and Singapore. Limited restrictions, based on already existing or amended legislation, as is the case in this country, are brought to bear in Australia, Brazil, France, Japan, the Netherlands, and the United Kingdom.

Some northern European countries have responded to public pressure to impose strict regulations by new legislation. Denmark, for example, prohibits the deliberate release of genetically engineered organisms without permission of the Minister of the Environment. Germany, under great pressure from the Green Party, which focuses on environmental issues, also enacted strict regulations in 1990.

It is safe to say that we are far from any universal standards for field trials. There is an international effort on the part of the European Community to reach a consensus. Member states were supposed to have drafted national laws by October 1991 concerning deliberate release of genetically modified organisms. Approximately one-half of the member states had done so by the deadline.

It is still unclear how diverse countries will balance the factors of industry lobbying, environmentalist pressures, scientific findings, and concerns about economic competitiveness in forging their biotechnology regulations. Certainly, with the arrival of "Europe '92" during a worldwide recession, EC members were not yet speaking with a common voice.

A long-term experiment going on in Britain may have far-reaching effects on future EC attitudes. PROSAMO, "Planned Release of Selected and Manipulated Organisms," is a four-year study, funded half by the British government and half by a consortium of interested companies, including DuPont (UK) Ltd., Monsanto Europe, and Plant Genetic Systems of Belgium.

Sixty-thousand seeds of canola, engineered to contain genes rendering them resistant to the herbicide Basta or to the antibiotic kanamycin, have been planted in three locations along with untreated canola. Researchers are looking to see if the treated plants are able to establish themselves as weeds in or outside the fields or transfer their transgenic attributes to wild plants.

So far no negative effects have been observed. This large-scale experiment will be followed by others using other species and genes. Eventually, though, according to the project's chief scientist, Michael Crawley, an ecologist at London's Imperial College, individual experiments on specific crops will have to give way to general principles governing the safety of environmental releases.

Willy de Greef, product manager at Plant Genetic Systems, expresses it bluntly. According to de Greef, if companies are forced to the prohibitive expense of field testing each and every crop or gene combination, "that will be the end of agricultural biotech."

FOOD—OR "FRANKENFOOD"?

There is presently no general scientific basis for presuming that foods from organisms into which new substances have been genetically engineered will be safe for human consumption. (Environmental Defense Fund, 1992)

If the public understood the technology, they would understand that part of their emotional reaction is irrational. (Dr. Susan K. Harlander, Professor of Food Science and Nutrition, University of Minnesota, 1992)

. . . plentiful, bioengineered foods threaten to upset the doomsday scenarios of famine, pollution, poverty, and ecological disaster to which so many of the antiscience groups are attached. (Elizabeth M. Whelan, President of the American Council on Science and Health, 1992)

In May 1992 the FDA announced that foods derived from genetically modified plant varieties would be judged on their characteristics, not on how the plant genes may have been manipulated. In keeping with the Bush–Quayle White House philosophy, the guidelines left it up to industry to determine if a new food presents sufficient risk to require federal oversight. They did not mandate labeling to inform customers or the FDA that a food product is the result of genetic manipulation. The ruling set off a flood of criticism from consumer, environmental, and food groups around the country.

Shortly after the announcement of the new FDA policy on genetically engineered foods, a cartoon in the *New Yorker* pictured an evil-looking scientist gleefully rubbing his hands over his crop of leafy plants which bore menacing faces. The caption read, "That's splendid news from the FDA, my pretties." A letter written by Paul Lewis of Newton

Center, Massachusetts, to the editor of *The New York Times* called food from transgenic plants "Frankenfood."

These are perhaps extreme reactions, but they were in response to what many considered an extreme policy decision. The May 1992 policy statement, which had actually had been made jointly by Vice President Dan Quayle, chairperson of the COC, and Louis Sullivan, Secretary of the Department of Health and Human Services, the FDA's parent agency, came under fire in other formats as well.

The FDA had both expected and solicited objections. The May announcement was actually a preliminary statement, which called for public input. By the end of the extended public comment period, the FDA had received three thousand written comments. Among the more vocal of the dissenters was the Pure Food Campaign, which we described earlier, led by activist–attorney Jeremy Rifkin.

Their ranks had swelled by 1993 to include over one thousand chefs, as well as other environmental and consumer organizations. The tactics of the PFC were aimed in particular at Calgene's Flavr Savr tomato, genetically engineered to resist rotting. The PFC claimed that these tomatoes might present a potential threat to consumers because the fruits had been given a bacteria-derived gene which renders those bacteria resistant to the antibiotic kanamycin. They warned that this might lead to a population of kanamycin-resistant bacteria in people who ate the tomatoes.

The gene is used in the process of making the transgenic tomatoes only as a "marker" gene, that is, one which makes it easier to pick out those plants which had been effectively transformed with the antirotting characteristic, by linking it to the resistance marker gene. Then the transformed plants can be selected out at a very early stage, rather than waiting until they bear fruit, because the plants would be more resistant to kanamycin than the nontransgenic plants.

Scientists supportive of the FDA argue that it would be extremely unlikely that any of the marker genes could survive a passage through a person's stomach, then move into bacteria in the intestines, and render the bacteria resistant to kanamycin—and have all of this happen in a person who needs or is using kanamycin. Calgene maintains that extensive testing has proven the safety of the tomato. Nevertheless, Calgene has asked the FDA to judge the tomato under the agency's most rigorous food additive requirements, in order to counter charges that the product has not been under close scrutiny.

Rifkin supporters, despite such assurances, regarded the risk as important enough to threaten a worldwide boycott of the products of the Campbell Soup Company, the New Jersey–based major food dealer. Campbell has financed the development of genetically engineered tomatoes at both Calgene and DNA Plant Technology of Cinnaminson, New Jersey, in return for various marketing rights.

The FDA did emphasize that it will demand approval and labeling if the genes used increase the concentration of a naturally occurring toxicant in the plant, introduce an allergy-causing substance, or significantly alter the concentration or availability of important nutrients for which the plant is ordinarily eaten. Those determinations are to be made by the producers, following guidelines set up by the FDA.

These latter assurances were met with skepticism from other of the policy's critics, alarmed by the notion of a self-policing industry. Jane Wessler, biotechnology specialist with the National Wildlife Federation, labeled the policy "outrageous." She said, "It is not protective to have a voluntary program for an industry that is just so competitive and trying so hard to get the first products to market."

Were supporters of the FDA's existing principles trying to make a case on their behalf, they might make it with something like the following summary (derived chiefly from government publications):

> Our diet already includes a wide variety of chemical substances, such as starch, sugars, proteins, vitamins, pigments, and flavorings. Hundreds of new varieties of food plants enter the market each year. For centuries, breeders have used all available techniques to develop advantageous combinations of genetic traits. Now that goal can be reached through genetic engineering.
>
> The United States has a safe, nutritious food supply. Foods derived from more traditional methods such as hybridization are not routinely subjected to scientific tests for safety. The developers and producers of our foods have a legal duty to ensure the safety of their products. In order to protect public health, however, the FDA will require testing new foods whenever there is a question about their safety. The FDA will also provide scientific guidance on how the foods should be evaluated.

This guidance will assist the food industry to monitor a variety of characteristics. If there are changes in the amount or availability of nutrients for which a plant is widely consumed, such as vitamin C in tomatoes, such alterations should be noted by labeling. The likelihood that gene insertions will activate pathways that will produce toxic substances in plants is low, but if such a possibility is suspected, the plants must be tested.

The producers are expected to monitor carefully the characteristics of the genes which have been inserted into the transgenic plants. The proteins made by the plant cells in response to the instructions in those genes generally are expected to be harmless, as are the thousands of different proteins that we now consume daily. True, many foods such as milk, eggs, fish, and nuts can produce an allergic reaction in some individuals. We have good analytical techniques to determine if such substances have been produced in new food varieties. We do have to admit that there is no practical method to predict if a new, untested protein would cause an allergic reaction in some people.

Changes in carbohydrates, fats, or oils which result in their becoming indigestible or toxic should be monitored and corrected. Although animal feeding trials of foods from new plant varieties are not routine, they could be used to ensure safety. However, in general, substances that have had a history of safe food use and closely related substances would not require extensive premarket safety testing. It would be impractical and impossible to try to protect people from any possible risk, no matter how unlikely.

Labeling any food which has been modified by genetic engineering would serve only to confuse potential consumers. There is so much misunderstanding about modern genetic science that such unnecessary labeling might lead to harmful boycotts. Even the American Medical Association has confidence in these products. In 1991 the AMA's Council on Scientific Affairs recommended that the AMA "Endorse or implement programs that will convince the public and government officials that genetic manipulation is not inherently hazardous and that the health and economic benefits of recombinant DNA technology *greatly exceed any risk posing to society*" [italics added].

Moreover, in 1990, the International Food Biotechnology Council, which includes representatives from 26 major food companies, issued a 400-page report, which recommended "no additional regulatory measures" for genetically engineered food products. It concluded that "existing laws and practices" provide a suitable framework for assessing risks.

And besides, the food industry has the technical capability and the will to monitor itself. After all, nobody has more incentives to bring safe and natural products to market than the food industry. Companies will use plants with a history of safe use and will gladly consult with FDA officials on a voluntary basis about forthcoming products.

Those opposed to the FDA guidelines might argue along these lines:

Putting genes derived from bacteria, unrelated plants, or even animals into crop plants is not just another step in the long tradition of modifying plants by cross-breeding. It is a totally new and unique method. Even the USDA admits that researchers cannot completely control the location where the selected gene inserts itself into the modified crop's genetic material. This could activate or deactivate genes in the plants and result in toxic substances undetectable unless the plants are carefully tested.

Genes direct the manufacture of proteins. Many of these can cause serious allergic reactions. The FDA may recommend precautions when companies engineer foods that contain known allergens, but what about all the new and possibly dangerous proteins that these plants will make?

And what about religious or personal objections to certain kinds of foods? Vegetarians and observant Jews, Moslems, and Buddhists need to know if animal genes have been spliced into plants. Food is more than just nutrition. It represents comfort and safety and graces our social rituals—weddings, family gatherings, celebrations, and times of mourning. So it is not surprising that thousands of people from all across the country have responded negatively to the FDA guidelines, which permit genetically engineered food to go to market without labeling and allows industry to decide when this food should be subject to safety testing.

Such labeling is necessary because we have a right to know what is in our food supply and how it was produced, and we must be allowed to make informed choices about what we are eating. We heard all the promises about the blessings of DDT and nuclear power, and when problems arose, neither industry nor the government was completely candid in reporting them. We are very skeptical about the food industry policing itself when its products stand to make enormous profits.

If industry wants us to trust in the safety of our foods, such confidence will not be gained by avoiding labeling. We agree with the reasonable, common-sense approach offered by the Environmental Defense Fund (EDF) in its 1991 position paper "A Mutable Feast: Assuring Food Safety in the Era of Genetic Engineering.

The EDF proposes that the new substances created in genetically engineered organisms be subject to the same premarket safety testing as are food additives. The EDF also asked for detailed labeling and a requirement that manufacturers notify the FDA at least 90 days before the new foods are marketed.

Even a congressional Office of Technology Assessment 1992 report stresses that agricultural biotechnology will never reach its vast promise until consumers' confidence is attained. The report maintains that a requirement to notify the FDA of the marketing of all these new foods, including data that the company used to determine that the product is safe, would not be an undue burden on the food industry.

The EDF rightly points out that the products of the expression of newly introduced genes should trigger regulatory scrutiny under the 1958 Food Additives Amendment to the Food, Drug, and Cosmetic Act. These regulations require labeling information about flavoring, preservatives, or dyes—why not new genes or their products? Besides, "organic foods" and irradiated foods are now labeled based on their product and processing techniques.

As far as the AMA is concerned, it can't have things both ways. The recommendations by its Council on Scientific Affairs also urges Congress and federal regulatory agencies to "develop appropriate guidelines that . . . will ensure that adequate safety precautions are enforced." We think that the FDA guidelines are inadequate for protecting the health of consumers and do not protect their right to informed consent.

As the arguments flowed back and forth, Calgene looked forward to marketing its Flavr Savr rot-resistant tomato sometime in 1993. It would be the first in what is expected to be a long line of genetically engineered fruits, vegetables, and other foodstuffs, which probably will be offered to consumers over the next several years. Company executives predicted that agbiotech would soon attain the status and scale that medical biotech reached in the late 1980s. In early 1993, Calgene awaited the expected final approval from the FDA before sales of the Flavr Savr tomato could begin.

Then, on January 11, 1993, the Campbell Soup Company announced that it had no immediate plans to market the Flavr Savr. Calgene officials maintained that their relationship with Campbell had not changed and that they were still proceeding on schedule with plans to introduce the Flavr Savr into North American markets in 1993. However, it seemed that, for the immediate future at least, the beleaguered tomato would not appear in Campbell's soups or sauces.

According to the January 12, 1993, *New York Times*, a Campbell spokesperson denied that the company was responding to boycott threats, maintaining that "Consumer campaigns have no effect on our planning process in this regard." On that same day, DNA Plant Technology announced that it had developed a marker gene derived from a plant rather than a bacterium.

As far as some other kinds of transgenic organisms are concerned, there is even less certainty about how they will be regulated. The EPA has not yet established rules for the amount of allowable pesticide residues in crops that have been altered to make their own pesticides, such as Bt. The USDA's Food Safety and Inspection Service is in the preliminary stages of preparing guidelines for transgenic livestock, but these are not expected to be available any time soon.

The FDA Office of Seafood has claimed responsibility for regulating transgenic fish and shellfish as foodstuffs when they are developed for that use. Currently, none of the researchers in the dozen or so U.S. laboratories genetically engineering fish plans to introduce them into open waters. Proponents argue that a few genes added to a fish species would have little environmental consequence. Others disagree, pointing out that adding traits to a fish that would extend its natural habitat—such as cold resistance—might allow the fish to range over new geographical areas, perhaps dramatically altering the ecology of a marine region. Whatever the case, compliance with federal regulations concerning the release of transgenic fish is now voluntary.

*   *   *

In the end, we are back to perspectives. Do we have anything to fear from our patchwork regulatory system, now evolving toward less supervision and more voluntary compliance in creating and marketing genetically engineered foods and releasing ("introducing") transgenic organisms into the environment?

Is there only a slight chance of creating serious problems, problems which we could overcome or live with anyway? Or are we to relive the history of the regulation of chemical hazards in the 1960s and 70s—first covered by public health statutes dating from the turn of the century? It took substantial damage to humans and to the environment before our government developed a new generation of laws, specific to the chemical industry. We are still paying dearly for decades of neglect, as we also try to cope with an annual production in this country of some 60 million tons of hazardous chemical wastes.

How much evidence is demanded to support one's position? Neither the biotechnology industry nor its severest critics can offer sufficient documented arguments to convince the other that we are on one hand facing a future bright with biotechnological solutions to our agricultural crises or on the other taking unacceptable risks with biotechnological quick fixes.

Certainly, some industry representatives are optimistic about how these arguments will be resolved. Roger Salquist, chairman and CEO of Calgene, was quoted in the *Chicago Tribune* in late 1992 as announcing that "the agbio train has finally left the station. [The] pipeline of genetically engineered agricultural products will overflow from 1995 on." He added, "The bulk of the American people will accept these products with great enthusiasm as long as they demonstrate consumer benefits."

Market Intelligence, a Mountain View, California, research firm was likewise optimistic after extensive 1992 interviews with marketing and technical experts. In its report, *Agricultural Biotechnology: Staggering Opportunities in World's Largest Market*, Market Intelligence predicted that sales of these products will grow to $1 billion in 1997 and to over $1.6 billion by 1998.

Renowned physicist Max Planck pointed out more than fifty years ago that major scientific innovations are rarely accepted by gradually winning over their opponents. Rather, he maintained, the opponents eventually die out and the succeeding generation grows up with the then-familiar ideas. Such a gradual demise of its opposition would be fatal for the ambitious dreams of the new biotechnology. The secrets of genetic change have been unlocked, and those who stand to profit are in an escalating competitive race. As competitors, industry welcomes clear and unambiguous regulations, but not regulations that demand lengthy and costly premarket testing. Prolonged public mistrust coupled with expensive case-by-case product scrutiny is not a climate for success.

There is yet another perspective. Given the promise of biotechnology to augment and enhance our food supply—a promise as yet largely unfulfilled—dare we erect what might be unnecessary roadblocks to testing and marketing its products? The AIDS epidemic presents a cautionary example. Suppose that recombinant DNA research in the 1970s had been banned altogether because of fears that never were realized. We probably would not even have been able to understand the workings of the responsible virus when it was discovered in the early 1980s or have the molecular tools to hope to develop a cure or vaccine.

Given the economic stakes and the tantalizing opportunities for some of our brightest scientific minds to create new, unique, and possibly valuable transgenic organisms, the prevailing perspective—i.e., that the risk of lost opportunities outweighs the perceived risks of proceeding—probably will maintain its firm hold.

The future of agricultural biotechnology as well as that of biotechnology in general in the rest of this century will no doubt be heavily influenced by the recommendations of the man elected to replace Vice President Dan Quayle. Albert Gore, former senator from Tennessee, was a leader in Congress in the fight to protect the global environment. Gore organized the first congressional hearings on toxic waste and helped lead the fight which culminated in the passage of the Superfund Law to clean up hazardous chemical dump sites.

His commitment to the health of the global environment should encourage those who would welcome more regulation of transgenic products. At the same time, as former chairperson of the Science, Space, and Technology Committee in the Senate, Gore played a pivotal role in promoting the use of agricultural biotechnology to help solve the problems of world hunger. Can these two philosophies live in harmony?

During his campaign, Vice President Gore, when asked to summarize his views on biotechnology responded: "Biotechnology is one of the critical technologies for the twenty-first century." He added: "We should bring the regulatory debate out of the back room of Vice President Quayle's Council on Competitiveness and into the [public] hearing rooms so that the public can become involved."

That seems already to have begun. Scarcely two days after he took office, Vice President Gore announced the abolition of the Council on Competitiveness. Gore promised to establish a process that would give the public much more information about proposed regulations, saying that "no longer will special interests receive special favors."

The future will reveal whether or not the USDA, which has an uncomfortable mandate both to promote and to regulate agricultural biotechnology, will be able adequately to "involve" the "public"—those who stand to be affected for better or worse by our government's economic and technical decisions.

Chapter 10

A WORLD OF GENES

Academics enjoy academic arguments. The favorite pastime in college and university faculty offices and meeting rooms is intellectual jousting—combat of the safest sort. Nothing is at stake for the winners or losers, except perhaps temporary embarrassment, but never banishment from the ivied ivory towers.

Outside of the hallowed halls, however, with the advent of the New Biology, the stakes for participants in debates over its promises and perils are much higher. We are all players in this game, as transgenic organisms are being created, released into the environment, and brought to the marketplace.

Our natural environment, teeming with organisms knit together in a complex web of interacting forms, sometimes competing, sometimes cooperating, is being asked to accept into its midst living creatures whose genetic material has suddenly been augmented with genes from other organisms. The creatures, moreover, derive from acts of genetic engineering rather than the age-old natural process of sexual reproduction.

Transgenic plants sport genes from plants with which they could never cross in the ordinary scheme of things, as well as genes from utterly unrelated creatures such as fish or bacteria.

Living organisms operate with a common currency, their genes. We now realize that not only are there many genes which are common to all forms of life, but that in principle, at least, virtually any gene from a given organism can be made to function in the cells of any other organism. We have already pointed out some of the early results of this unprecedented knowledge and its application, genetic engineering.

Both the knowledge and the technology are so new, and advancing so rapidly, that no one can authoritatively or clearly enumerate and evaluate the rewards and risks. We have already seen some of the promises of transgenic plants unmasked in the case of the Bt insecticide–producing crops. The anticipated widespread use of these may trigger a population explosion of Bt-resistant insects. We also have chronicled the continuing controversies over field testing and genetically engineered foods.

The promise of rewards alongside the warnings of risks in the latter controversies display widely differing perspectives. Some have hailed the arrival of genetically transformed plants as timely contributors to a better world and urge relaxation of restrictions on their use. Others vigorously contest the practice of releasing such plants into an environment unprepared for what they consider may be irreversible and harmful consequences. Still others are optimistic proponents of genetic technology if it is used with what they consider to be deliberate caution.

To most people, caught somewhere in the midst of trying to interpret the charges and countercharges, the decisions made by governments, industry, or farmers are anything but academic. Water, soil, and food are shared by all, whether by decision-makers or by observers.

An analysis of the disagreements over these issues, installments of which appear on an almost daily basis in our newspapers and popular magazines, as well as in scientific journals, reveals that they are often much more than a dispassionate weighing of evidence. Decisions are not made in the manner of a judge interpreting statutes. They are often made in the context of a "world view," that is, how they fit into one's philosophy of how we humans should act if we are concerned about the welfare of the rest of the living world and the environment that supports us all.

As a concrete example, let's take another ongoing, active controversy whose future resolution will have important consequences and whose most active participants are not disinterested academics, but people whose basic attitudes toward science and technology lead them in quite opposite directions. While some promise us a better world, others warn of a "bitter harvest."

THE WEED KILLERS

Modern farming, in this country at least, has emphasized maximum productivity to the point at which only 2 percent of our population raises the food for the other 98 percent. This predominately large-scale agribusiness approach is a very recent development in human history. Traditionally, farms were small and labor-intensive and supported a wide range of plant and animal species. These farm crops were genetically diverse—and very low-yielding by today's standards. Now, farming has joined the ranks of human activities that threaten the long-term stability of the Earth's environment.

Out of necessity we now put a premium on output of crops. Millions of acres of genetically homogeneous plant varieties are force-fed a diet of chemical fertilizers, drenched with doses of herbicides and other pesticides, and then harvested from irreplaceable soils, which are often eroding at an alarming rate. Farming demands massive amounts of petroleum products, further draining our supply of nonrenewable fossil fuels. Conflicting with the other water needs of growing urbanization, demands for increased irrigation are being put on our groundwater supplies at the same time as its very sources are sometimes being contaminated by farming chemicals.

We will never return to the days when each family or locale raised its own food, any more than we again will all make our own tools or clothes.

But can we use biotechnology to enhance increased agricultural productivity while simultaneously sharply reducing the degradation of the natural environment brought about through mega-farming?

Certainly, weed control is essential for all crops. Weeds are major pests, whose unchecked growth competes for space and nutrients, drastically reducing crop yields and farmers' profits. Until the middle of this century, farmers controlled weeds by laboriously tilling the soil and rotating crops. After World War II, the synthetic chemical industry, having invested heavily in research on chemical warfare and eradication of insects carrying malaria, turned to the alluring market of agricultural chemical pest control.

The production of agricultural pesticides—killers of insects, fungi, nematodes (microscopic worms), and rodents as well as weeds—had an overwhelming impact on farming practices. Labor-intensive crop rotation and plowing gave way to herbicides, potent chemical weed-killers. Town, county, state, and federal agencies and even private citizens turned to herbicides to beautify lawns, parks, and forests.

Now, more than 300,000 tons of herbicides, mostly for agriculture, are spread annually over the soils of this country. They account for 60 percent by weight of all pesticides used. Over 95 percent of all major row crops grown in this country have come to depend on chemical herbicides to remove competing weeds.

One reason that herbicide dependence has become so firmly entrenched is that farming is a business—and the use of chemicals is now many times cheaper per unit of production than the use of expensive mechanical equipment, which consumes fuel and requires people to be paid to operate it.

It is true that in contrast to the variety of chemical herbicides now available, which kill a wide spectrum of weeds, there are a few "natural" weed killers on the market. Most are fungi which have been shown to attack specific weeds quite effectively. Others are insects which feed on certain weed types as well. Such biological control approaches are still in the early stage of development, and so appear to be far from competitive with the widely used chemical herbicides.

The term chemical, given the horrendous history of industry-generated chemical pollution over the last 50 years, understandably tends to conjure up negative images. Although all forms of life are, in the final, reductionist analysis, complex arrangements of chemicals, we often tend to associate chemistry with the unpleasant smells of a reluctantly remembered chemistry course or the toxic wastes for which we continue to pay dearly, both in money and in environmental degradation.

Industry has brought us lifesaving antibiotics, medications, pharmaceuticals, and a host of other products—chemicals all—which enhance our human condition. But do herbicides belong in that benign category?

The power of herbicides was amply demonstrated during the Vietnam War. More than 120 million pounds of herbicides were used to drench the forests, plantations, and rice paddies of South Vietnam, an area about the size of Florida. After the war, every ecosystem in Vietnam had been seriously damaged, some even destroyed beyond hope of repair.

Farmers are more selective, ordinarily dispensing herbicides in measured amounts for maximum effectiveness. Despite their precautions, however, according to some studies, herbicide use has had a significant public health impact. A 1987 National Research Council report, *Regulating Pesticide Residues in Food: the Delaney Paradox*, studying the cancer-causing risk of pesticides on food estimated that herbicide residues can pose a slight risk of causing benign and malignant tumors in consumers.

While it has been shown that many herbicides are not acutely toxic to humans, there has been much concern over the possible dangers of chronic toxicity, i.e., disorders caused by low-level exposure over a prolonged period. Relatively little research has been done on the chronic health effects of widespread herbicide use on the general population. Reports by the National Research Council in 1987 and the Environmental Protection Agency in 1985, among others, have linked various herbicides with cancer, birth defects, central nervous system disorders, behavioral changes, and skin diseases.

The EPA cited especially strong evidence linking alachlor, which along with atrazine is the most frequently applied herbicide, to malignant tumors in humans, according to a 1985 Office of Drinking Water analysis, even though the use of both is subject to safety regulations. The EPA regards atrazine as well as a possible human carcinogen. In 1988 the EPA reported that 74 pesticides had been detected in the groundwater of 38 states, among which were 21 herbicides, including alachlor and atrazine.

We do not yet have very much information on the problems that might be caused by the exposure of farmers and farm workers to pesticides. In 1989 the EPA canceled registration of some herbicides containing bromoxynil, which causes birth defects in some laboratory animals and may pose risks for those who use it. The agency now requires workers who load, mix, or apply this chemical to wear protective garb. Humans are not the only ones at possible risk. A University of Missouri study has described bromoxynil as "highly toxic" to bluegill fish and "extremely toxic" to channel catfish.

A 1992 study carried out by the National Cancer Institute concluded that pesticides are a likely cause of some forms of cancers such as Hodgkin's disease, multiple myeloma, and leukemia, found more frequently in farmers than in the general population. The study suggested that pesticides may be interfering with the farmers' immune systems, allowing tumors to gain a foothold.

Does the New Biology offer a way out of the chemical "fix" that is so much a part of modern farming? In Chapter 9 we explored the biotechnological approach to making plants which produce their own Bt pesticide. In Chapter 8 we described several successful attempts to create transgenic plants which are herbicide-tolerant, which means they can survive and prosper even if exposed to herbicide formulations or concentrations which would kill intolerant plants.

Certain plants are naturally tolerant to particular herbicides. For example, corn, wheat, oats, and rice tolerate certain herbicides that kill most serious weeds. However, crops are often sensitive to particular herbicides, or if not, the crops that subsequently will be planted in that same field are harmed by them. Having herbicide-tolerant crops would remove these limitations. In the words of the Congressional Office of Technology Assessment, such plants "are designed to tolerate higher levels or more potent doses of herbicides than nontolerant crops."

Research and development aimed at making herbicide-tolerant plants is high on the agenda of agricultural genetic engineers, consuming an impressive 30–50 percent of the industry's resources. As a matter of fact, the eight largest pesticide companies—Bayer, Ciba–Geigy, ICI,

Rhone–Poulenc, Dow/Elanco, Monsanto, Hoechst, and DuPont—are all active players in the race to develop herbicide-tolerant crops and bring them to market.

At least 20 other corporations have joined in the quest. Almost all major food crops, including soybeans, wheat, corn, sorghum, rice, and potatoes, as well as carrots, cotton, tobacco, tomatoes, and alfalfa, among others, are being urged to accept genes which allow the plants to defy the otherwise withering effects of chemical herbicides. These companies include many of the world's principal seed corporations, such as Pioneer Hi-Bred, Sandoz–Hilleshog, and Upjohn.

In 1989 Calgene was granted permission to field test its cotton variety, which has been engineered to withstand the weed-killer bromoxynil, made by Rhone–Poulenc. Cotton farmers now spend $150 million annually on herbicides, but still lose 7 percent of their crop to weeds. By switching to bromoxynil, Calgene estimates that the crop loss could be reduced to 2 percent or less.

By now, Calgene and others have already field tested, among others, soybeans, alfalfa, and corn, resistant, respectively, to glyphosate (Roundup), glufosinate (Basta), and sulfonylurea (Classic, Express). These plants are expected to go on the market by the mid-1990s.

Why has there been such a heavy emphasis on the development, testing, and marketing of crop plants which have had genes from plants, bacteria, or viruses added to their genomes, granting them greater immunity to specific herbicides? That is decidedly a nonacademic question. Rather, it is one whose ramifications reach into the lives and futures of us all.

One obvious motivation is profit. Industry analysts have predicted that the annual value of herbicide-tolerant seed could be as high as $2.1 billion by the year 2000. For example, it is estimated that a glyphosate-resistant soybean could increase the sales of the glyphosate-based herbicide Roundup by $150 million, while resistant canola could increase sales by several hundred million dollars during the life of Monsanto's patent on glyphosate compounds.

Industry spokespersons and other proponents have been quick to claim that the real value of these novel plants is that they will permit the replacement of older, dangerous pesticides with newer, more "environmentally benign" chemicals. They claim that the latter will break down more rapidly and be less toxic. They argue as well that these new products can be used at lower application rates than the older herbicides. Farmers will be able to make environmentally sound decisions about their herbicide use. Even soil erosion will be slowed, they say, because some farmers would be able to avoid excessive mechanical tilling of weeds, which leads to erosion. Instead, they could apply herbicides and plant impervious crops, while practicing no-till agriculture.

What could be more in tune with American ideals than profit and progress? Again, that depends on one's perspective. Probably the most widely read account of the still ongoing imbroglio over these herbicide-tolerant plants is the March 1990 report of the Biotechnology Working Group, *Biotechnology's Bitter Harvest*. This organization is a coalition of representatives of public interest organizations and numerous other activists concerned with biotechnology-related issues and their social and environmental impact. Their members include Margaret Mellon and Jane Rissler of the National Wildlife Federation and Nachama Wilker of the Council for Responsible Genetics.

Their report crystallizes a fundamental opposition to herbicide-tolerant plants. On the one hand, it underlines the known and possible deleterious effects of herbicides on organisms and the environment. The writers claim that "widespread use of herbicide-tolerant plants will increase the likelihood that additional residues of particular herbicides will contaminate food. . . . There remains considerable uncertainty as to whether the tolerance levels set by the Environmental Protection Agency protect public health." They cite the USDA's acknowledgment that "these new [herbicide-tolerant] crop varieties might carry more herbicide residues. . . ." That assumes, of course, that farmers, in their war against weeds, would apply more herbicides to these varieties than they would to ordinary plants, because these altered crop plants would not be damaged.

Further, they emphasize that since gene insertion is usually random, herbicide-tolerance genes could disrupt the functioning of other genes, so that "some genetically engineered plants might be less nutritious or even produce higher levels of natural plant toxins than nonengineered plants."

The report points to studies that reveal that more than 50 kinds of weeds have developed resistance to atrazine and others have recently evolved resistance to the new generation of low-dose herbicides such as the sulfonylureas. *Biotechnology's Bitter Harvest* warns that "Widespread use of herbicide-tolerant crops—with their associated potent herbicides—will exert significant pressure on additional populations of weeds to develop resistance to the herbicides." This, as we have pointed out earlier, could conceivably come about by the preferential survival and reproduction of naturally resistant weeds as well as possibly by cross-pollination from herbicide-tolerant crops to closely related weeds.

Widespread evolution of resistant weeds could make them poor candidates to replace older herbicides. Some pesticide experts have already recommended that sulfonylurea be mixed with the older, more toxic herbicides. And as far as soil erosion is concerned, the Biotechnology Working Group claims that there is a broad range of erosion control practices such as cover crops, contour farming, terracing, and crop rotation that allow minimal use of herbicides.

In short, "To those with high hopes for the environmental benefits from biotechnology," the authors of *Bitter Harvest* write, "herbicide-tolerant crops are at best a distressing misstep, at worst a cynical marketing strategy." In response to a press conference highlighting the release of the report, held outside the greenhouses of Calgene, Inc., Calgene chairperson Roger H. Salquist criticized the report as "pandering to the ignorance of the public" about agricultural biotechnology research.

Will the growing list of herbicide-tolerant crops being field tested and groomed for market lead to increased use of particular herbicides that will do further harm to organisms and the environment? Or rather, will fewer, more benign chemicals be employed as a result of these technological creations? That remains to be seen. What is certain is that the controversy over herbicide-tolerant plants is a prime example of how the rhetoric in the debate over a specific application of biotechnology reveals more than a disagreement about the niceties of a particular technique. The real debate is as much about a mind-set as about a methodology.

It is easy to oversimplify this debate and others over the use or abuse of the New Biology. However, there are two distinct attitudes which have revealed themselves consistently in books, articles, symposia, and speeches over the last several years. There has evolved, in fact, a confrontation that promises to be reenacted with increasing frequency until (or if) there is some resolution.

The first attitude is that of those who might be called technological optimists, represented by, among others, the biotechnology industry. In their view, nature is seen as something that can be analyzed to the point where its working principles can be understood. This knowledge, while never perfect, can lead to manipulation of nature, in this case through genetic engineering, so that nature is brought under control for purposes of our health and comfort. They admit that there are risks involved, but maintain that science can usually quantify and control them so that they at least become acceptable risks. Because of their confidence in their expertise and experience, the technological optimists would place the burden of proof that grave harm might result from, for example, herbicide-tolerant plants, on those who question the creation and use of such plants.

The opposing attitude is one which is held by people who often are labeled, even by themselves, as environmentalists. One sometimes hesitates to use that term, however, because of the image that it conjures up in the minds of many who might otherwise be supportive of environmentalist aims. The 1960s and 1970s saw an awakening of the realization that rampant consumerism had led to pollution of our air, waters, and soils. Our government responded by legislation that began to limit and reverse the previously almost uninhibited license to despoil our environment, from filling wetlands to dumping toxic chemicals into waterways.

Concurrently, society was changing drastically in other ways. A new sense of social freedom, which often overstepped traditional boundaries of public and private behavior, flowered in these decades. The Vietnam War exacerbated the divisions in our society, and by the time the 1970s were drawing to a close, the often overwhelming events which had destroyed forever the complacency of the 1950s had contributed to a distorted portrait of the environmentalist perspective. Highly vocal enthusiasts clamoring for increased protection for fragile, endangered habitats projected an unattractive image to many Americans.

The term came to be associated with obstructionists, those who stand in the way of progress. Many agreed that we should protect our environment, but was the life of a particular, allegedly obscure species of animal or plant, or the preservation of large tracts of wilderness which might otherwise be mined for minerals or tapped for oil, worth limiting our exploitation of nature for the necessities of our growing population?

Environmentalism seemed to many to be an amalgam of those distant days of protest, a movement with some laudable ideals but extremist in its opposition. Of course, while one person's extremism is another's zealous defense of principle, there are generally accepted limits which govern even heated debates. Not so for the group of Dutch activists who slipped into a test field of genetically engineered corn in Rotterdam in August 1992 and cut the stems of 900 of the herbicide-resistant plants. The damage cost the plant breeding company VanDerHave over $1 million and set its research program back at least one year.

Actually, in the highly vocal but ordinarily nondestructive debates over biotechnology, many self-proclaimed environmentalists quite readily concede its potential for positive contributions. In the words of Margaret Mellon, for example,

> . . . environmentalists are practical. They understand the need to exploit nature in order to sustain human life and the positive role genetic improvement has played in developing plants, animals, and microbes for food, fiber, and drugs. They do not doubt that the new techniques of biotechnology could provide important benefits in the future.

Michael W. Fox, Vice President of the Humane Society of the United States and an outspoken critic of transgenic animal biotechnology, claims that he is not in principle opposed to biotechnology. He says that "We need not abandon biotechnology as some evil or Promethean curse, but we must use it prudently, with respect, humility, and compassion."

However, environmentalists would place the burden of proof that widespread use of herbicide-tolerant plants would not lead to unacceptable harm on those who propose to make and sell these plants. Their position is that what appear to be the risks, for example, of increased use of herbicides or transfer of resistance to weeds, are unacceptable risks. Technological optimists deny that there will be increased use and contend that the spread of genes conferring herbicide tolerance can be controlled to the point where that risk is acceptable.

One difficulty often faced by environmentalists in defending their position is their evocative language, which often ignites the residues of resentment against them. Michael Fox, for example, in his 1992 book, *Superpigs and Wondercorn: The Brave New World of Biotechnology and Where It All May Lead*, speaks of developing a new attitude "toward the whole of creation and the creative process itself." Later he describes "nature's creative process" and asks for a "resacralization of nature." Margaret Mellon, writing in the 1991 collection of essays *The Genetic Revolution: Scientific Prospects and Public Perceptions*, describes biotechnology as having the power to "subdue nature" and defends the "values" of the environmental "movement."

These are terms that tend often to move the debate beyond science and into philosophy. They ask for a sea change in society, a holistic approach which takes into account not just immediate technological solutions for specific problems, but a wider view, sensitive to the possible long-term effects of technology on organisms and the environment. Environmentalists call for a deeper appreciation for the complex, often only vaguely understood interrelationships among living organisms and between those creatures and their environment.

Most importantly, they stress that any biotechnological process or product must take fully into account the fact that humans, because of our extraordinary intellectual abilities and potential for harm, while truly an integral part of nature, are also obliged to exercise care and protection in their activities, so that they do not do irreparable harm.

Although discredited by critics as having impractical, metaphysical elements, whether or not such a philosophy prevails will, in the view of environmentalists, determine not only the future direction of biotechnology, but the quality of life on the Earth for all future generations. It may yet prevail, but certainly at this point it can hardly be described as a widely accepted paradigm.

SUSTAINABLE AGRICULTURE

So the debate over herbicide-tolerant plants leads one into a larger context. It is one in which technological optimists and environmentalists struggle, not over one or the other technique in modern farming, but over the question of what kind of agriculture can give us adequate food supplies now and in the future while preserving a healthy farming environment for our descendants. This kind of productive, renewable system is known as sustainable farming. The New Biology has the potential to be an integral part of its growth and development—or to stand in its way.

Sustainable agriculture has at various times been described as *natural*, *organic*, *eco-agricultural*, *regenerative*, or *alternative*, among other terms. These are also evocative words. Some conjure up the image of a "back to nature" movement, perhaps by vegetarians, or at any rate, impractical dreamers. It is true that adherents of sustainable agriculture encompass a wide range of enthusiasms. For example, early support came from a desire to preserve small and moderate-size farms, on which farmers could adapt to local conditions of soil and water and thereby not only provide a more careful husbandry of the environment but sustain healthy rural communities. Put simply, good farming is, according to Wendell Berry, noted Kentucky poet, essayist, and farmer, "farming that does not destroy either farmland or farm people."

Others extend the goals of sustainable agriculture to include the amelioration of the exploitation of the labor of women and people of color in large-scale megabusiness agriculture. However broadly one conceives of sustainable agriculture, there seems to be a common denominator, which is clearly expressed by a definition offered by the Food and Agricultural Organization of the United Nations:

> Sustainable agriculture should involve the successful management of resources for agriculture to satisfy human needs while maintaining or enhancing the quality of the environment and conserving natural resources.

Over the last decade, the notion of sustainable agriculture has shown a remarkable transition from a fringe movement to a mainstream philosophy. In most agricultural colleges there are now programs committed to its study and development. Prominent examples are the Center for Integrated Agricultural Systems at the University of Wisconsin and the Leopold Center at Iowa State University. Two journals are devoted wholly to the subject, *The American Journal of Alternative Agriculture* and the *Journal of Sustainable Agriculture*.

In 1989 the National Research Council of the National Academy of Sciences issued *Alternative Agriculture*, a report that went a long way toward debunking one of the principal criticisms of sustainable systems, i.e., that they are not profitable. The report stated that "Farmers successfully adopting these systems generally derive significant sustained economic and environmental benefits." This NRC report emphasized the technicalities of sustainable farming and did not attempt to comment on the important social issues of labor and family farms. It underlined the notion of replacing the use of herbicides, as well as other chemicals, for the long-term protection of the environment. This principle is at the center of alternative (sustainable) agriculture.

Earlier we listed some of the options for weed control, which include modification of tilling practices, crop rotation, or planting cover crops such as rye or barley that inhibit certain weeds. All of these techniques as well as others require research and development on a scale far beyond the current levels. According to the U.S. General Accounting Office, only about 2 percent of the 2.2 million farms in this country have made the switch to chemical-free agriculture. The authors of *Biotechnology's Bitter Harvest* urge that taxpayer funding now supporting herbicide-tolerance studies be redirected to new approaches to weed management. They contend, as do others, that widespread adoption of herbicide-tolerant crops would "perpetuate the high chemical dependence of conventional farming."

Certainly, at least some facets of alternative agriculture have moved from the fringes of the farming community to the mainstream of serious consideration. This has been an important qualitative if not quantitative victory for its proponents. But does the notion of sustainable agriculture by its nature preclude a role for biotechnology? Not at all, at least according to many of those in attendance at the 1989 National Agricultural Biotechnology Council (NABC) Symposium on Biotechnology and Sustainable Agriculture. The NABC provides an open forum for participants from all sectors of science, business, and academia to discuss issues related to the impact of biotechnology on agriculture.

At this meeting, Robert M. Goodman, Executive Vice President of Research and Development at Calgene, in his keynote address co-opted the language of the environmentalists in trying to support his contention that "genetic manipulation is the proven environmentally safe way to address production challenges—both economic and environmental." He spoke of the need for "ecological stability," "ecological values," and emphasis on "stewardship." He promised that "some herbicide tolerances may result in lower overall uses of herbicidal chemicals and lower input costs for growers." To support this point Goodman cited Calgene's glyphosate-tolerant tomato, whose cultivation he predicted would lead to "significant decreases in overall herbicide usage."

The other keynote speaker was Charles Hassebrook, from the Technological World Agricultural Program at the Center for Rural Affairs in Walthill, Nebraska. Hassebrook agreed that biotechnology might be useful, for example, in finding new uses for some crops grown in sustainable systems or for creating crops resistant to diseases. However, he insisted that "the development of herbicide-tolerant crop varieties will continue the trend of making farmers more dependent on chemicals for weed control." He pointed out that if, for example, a corn variety were tolerant to the herbicide "Roundup," which kills just about any plant on contact, it would logically lead to relying totally on this chemical for weed control, eliminating mechanical weed control. This would also make it possible "for one person to farm more acres and for fewer people to farm the nation's land."

Will the imminent availability of the many varieties of herbicide-tolerant crops now under development and testing support the goals and ideals of sustainable agriculture? Despite the optimistic projections of the agrichemical industry and biotechnologists, it looks as though a significant shift will have to occur in conventional agriculture if the technological optimists and the environmentalists are to work as partners.

Joan Dye Gussow, professor of Nutrition and Education at Columbia University in New York City, points out in her insightful 1991 book *Chicken Little, Tomato Sauce, and Agriculture: Who Will Produce Tomor-*

row's Food? that conventional agriculture moves in quite the opposite direction. The bottom line is "efficiency" and "lowest cost" (not, we should add, necessarily incompatible with lowered dependence on chemicals, according to advocates of alternative agriculture). Much of our food production depends on using land as raw materials to be "turned into 'food' by modern industrial processing." She emphasizes that "fertilizers based on fossil fuels cannot be used forever. Soil damage by misuse of chemicals . . . aimed at ever-increasing yields cannot produce forever. Water damaged by chemical runoff or leaching cannot support agriculture in the long run."

This amalgam of economic and environmental issues has led us into a dilemma in which the need for increased productivity has been met by practices which are counterproductive in the long term. It is a problem that "you couldn't produce your way out of," according to Professor Dennis Keeney, Director of the Leopold Center for Sustainable Agriculture at Iowa State University.

At that same NABC conference, Iowa State University philosopher Gary Comstock, while admitting that herbicide-resistance technology is theoretically compatible with the USDA definition of sustainable agriculture, concluded that the two are not compatible in fact. Again he pointed to the dilemma faced by the American farmer who cannot stay in business without high yields. This leads to a cycle of fertilizers, herbicides, and other pesticides. Farmers who enroll in government subsidy programs are driven toward "large combines, large amounts of capital, large fields, and tons of purchased inputs."

Despite the fact that the 1989 NRC report argues against what has long appeared to be a conflict between the notion of sustainable farming practices and sufficient productivity, it appears that the average American farmer is far from ready to make a major transition from chemical dependence to low-chemical-input crop production. But even if growing herbicide-tolerant plants does turn out to lead in that direction, there seems to be little prospect in the foreseeable future for achieving the wider vision of sustainability, one based on smaller farms, more adaptive to local conditions of climate and soil and less polluting and embodying a new social and economic organization of agriculture, which aims at resolving inequalities of class, gender, race, and ethnicity. Biotechnology is a tool, not a magic wand.

AGRICULTURAL BIOTECHNOLOGY AND
THE THIRD WORLD

In 1991 a series of detailed essays appeared in a special issue of the British journal *Science as Culture*. Several writers explored various aspects of this wider vision of farming. The lead editorial frames the journal's view of the situation bluntly:

> Encouraged by government subsidy, . . . Western countries produce food surpluses, whose low price export to Third World countries further undermines those people's ability to produce their own food. Moreover, intensive methods have been extended to large parts of the Third World, where cash crops for export have further replaced local subsistence crops. Thus the entire agricultural system has come to epitomize an epochal industrial malaise, having misguidedly sought total control over nature and maximum productivity of the land.

What is the relationship between the New Biology about to be offered to farmers in the form of genetically engineered crops and the world outside the industrialized developed nations? Does it offer new hope to the Third World, or might it make the gap between the affluent and the poor even wider?

In predicting what role agricultural biotechnology will play in the formidable challenges faced by the less developed nations of the world, one runs the risk of theorizing too simplistically about a situation of great complexity. Any substantial change in farming practices, as well as any other important social change, depends not only on new technologies but just as importantly on interactions within the cultural, political, economic, religious, and military milieu of the region in question.

Throughout Africa, for example, the ruinous depletion of forests and water is a legacy of land distribution patterns dating from the colonial era. Commercial farms have taken the best land, driving millions of the poor to drier, less fertile areas. The World Bank estimates that five million acres of forest are cleared each day in Africa in a quest for land to plant, as well as for fuel and lumber.

According to the 1990 edition of the annual *State of the World*, a respected annual global environmental overview prepared by the Worldwatch Institute, "Economics and national and international policies will determine whether biotechnology can achieve its technical potential to provide more food for a hungry world." The editors go on to point out that, in their opinion, despite promises of miracle crops that will help to feed the hungry millions,

> Corporations will use biotechnology to develop crops and agricultural products they expect to be profitable, and, if possible, which they can patent. . . . resistance to herbicides—which will, in fact increase the use of these chemicals—is receiving R&D priority.

There are quite different opinions. In a speech delivered to the Second Annual American Society of Microbiology Conference on Biotechnology in June 1987, Monsanto's Howard A. Schneiderman predicted, "If biotechnology is effectively used by developing countries, it can help them secure the sustainable agricultural productivity they need to feed and maintain themselves." The major seed company ICI Seeds released a publication in 1989, *Feeding the World*, in which it was argued that biotechnology "will be the most reliable and environmentally acceptable way to secure the world's food supplies."

In 1990 Ellen Messer and Peter Heywood, both associated with the World Hunger Program at Brown University in Providence, Rhode Island, reviewed the state of plant biotechnology, current and projected, in their carefully documented report *Trying Technology: Neither Sure nor Soon*. They point out that we can now realistically manipulate only a handful of genes for useful characteristics—for example, herbicide tolerance or virus and insect resistance. Other desirable traits, such as tolerance to temperature extremes, drought, or salinity, are governed by a complex of genes and will be much more difficult to genetically engineer.

They predict that agricultural biotechnologies will probably have little impact during the rest of this century. Over the next century,

however, they do see them as having considerable impact on the allevia-
tion of hunger. That assumes that more and more useful transgenic crops
will be created, "if the private firms, or the scientists who lead R&D
programs, make producing more and better quality food, at lower cost, a
priority for the utilization of their research findings."

The vision of miracle crops entering the fray against world hunger is
alluring, and the prospect is certainly not to be discounted. Meanwhile,
the skills and ingenuity of biotechnologists here at home are beginning to
create products that literally could put thousands of Third World farmers
out of business and cost their countries millions of dollars in agricultural
exports.

For example, the United States imported $300 million of coconut,
palm, and palm kernel oil in 1991–1992 for use in food, detergents,
soups, and shampoos. More than nine million people in the Philippines,
Indonesia, and other Asian countries grow, harvest, and process coconuts
for this export trade. Now at Calgene, scientists have genetically altered
the rapeseed plant to produce a canola oil rich in lauric acid, enough to
make the oil a possible substitute for the tropical imports.

Rapeseed is already grown in six U.S. states and Canada for food
and other uses. Calgene is now field testing the new plants in Michigan,
and plans to have its subsidiary in Georgia, Ameri-Can Pedigreed Seed
Co., produce and sell the modified oils to industrial users. Meanwhile,
Calgene is testing yet another genetically altered variety of rapeseed that
is richer than normal in stearic acid, which makes the oil low in saturated
fat. This would be useful in margarine and in chocolates and other
confections—and take over part of the $2.6 billion market for cocoa
butter, the second-largest tropical commodity after coffee.

DNA Plant Technology Corp. of Cinnaminson, New Jersey, has
managed to introduce genes into rapeseed as well. These genes reduce
the levels of saturated fat, making the oils less apt to raise the cholesterol
levels of consumers. The oils are also odorless and smokeless at high
temperatures, making them ideal for cooking oils.

The U.S. annually imports about 400 tons of vanilla from Mada-
gascar, one of the poorest countries in the world. A vanilla is now being
made in huge vats of vanilla orchid cells in the laboratories of Escage-
netics Corp., in San Carlos, California.

It certainly makes good business sense to gain control over the production of a substance and assure its availability, free from political and social sources of price fluctuation. But what of the effects on the struggling suppliers in the Third World? Some industry spokespeople actually forecast a positive impact. They predict that having a stabilized supply of oils and flavors will increase demand and thereby open new markets for Third World farmers.

Others are less optimistic. According to Martin Kenney, professor and rural sociologist at the University of California at Davis, "Overall, the Third World countries will probably be the losers. . . . the technology is not going in the direction that would benefit them." We find ourselves once again heading toward an uncertain future, painted in quite different colors by optimism and pessimism about the benefits and drawbacks of the New Biology.

If gene-altered crops are someday to serve the best interests of the less developed nations, tomorrow's agriculture will probably have to be guided in new directions by institutions already in place. The *State of the World: 1990* report urged increased support for the international agricultural research centers, which for several decades have been at the forefront of efforts to bring the fruits of modern agricultural science into increased food production in developing countries.

There are now 17 international agricultural research centers, and another dozen or so centers that are working in concert with them or are being set up. They have evolved beyond their original mandate to increase food production to an additional goal of sustainable agriculture. These centers are now committed to searching for alternatives to reliance on fertilizers and pesticides while keeping crop production rising.

There are now some 40 donors supporting the centers, which are under the leadership of the Consultative Group on International Agricultural Research (CGIAR). These include the World Bank, the UN Food and Agriculture Organization, the UN Development Program, and the Rockefeller and Ford foundations. A modest annual budget of $300 million aids in the support of 1,700 senior scientists among a total staff of some 15,000. A major study in 1988–90 by the World Bank and other agencies concluded that many potential benefits could accrue from integrating modern biotechnology with conventional agricultural research. They cited as significant public-sector investments in so-called "orphan commodities" such as cassava, sorghum, or sweet potatoes, crops not traditionally of interest to industrialized countries but of major importance to the producers and consumers of the Third World.

The CGIAR, because of its ubiquitous presence and 20-year history of experience in establishing relationships with governments and local agricultural agencies, might seem to be a logical conduit for whatever agricultural biotechnological advances might be available to play a useful role in Third World farming. That may happen, in fact, on a large scale eventually, but at this point the organization is under the constraints of the considerable financial uncertainty of its funding.

Its involvement with biotechnological research is now modest and is predicted to remain so. However, it is implementing its established network system to pass on useful research techniques to national systems. For example, it is working with the Cassava Biotechnology Network, supported by the Government of the Netherlands, as well as with a consortium of laboratories in Japan, Europe, the United States, and Europe which are assisting two centers: The International Rice Research Institute in the Philippines and the Centro Internacional de Agricultura Tropica in Colombia.

At this point the reader may have the impression that even if the plant genes put into service by biotechnologists do give the industrialized countries of Europe and North America secure supplies of commodities traditionally purchased from Asia, Africa, and Latin America, increased public and private support for technology transfer by the CGIAR and other international organizations will make up for what will be a temporary inconvenience. Perhaps new "miracle" crop varieties engineered for specific local conditions will offer the poorer nations abundant food and salable commodities.

That rosy scenario is at this point an idealist's dream, however. As a matter of fact, critics warn that the creation and marketing of the gene-altered plants of the future, rather than benefiting the Third World, may instead perpetuate the injustices of colonialism. Why? Because the currency of the New Biology is genes. The genetic information encoded in these genes is the crucial natural resource needed by biotechnologists. Crop plant genes are conveniently stored in seeds—which may now justifiably be regarded as the world's most precious raw materials.

Most of the abundance and diversity of that precious lode of genes is found in the Third World countries. Seeds have traditionally been regarded as the "common heritage of mankind" rather than a resource belonging to the indigenous peoples in whose lands these genetic treasures are hunted. The genetic riches of the less-developed nations have been systematically extracted for many years. And now, the very habitats in which the seeds are found are disappearing, victims of the effects of burgeoning population, poverty, and war.

SAVING THE SEEDS

> The greatest service which can be rendered to any country is to add a useful plant to its culture. (Thomas Jefferson)

Of the approximately 250,000 species of flowering plants that have existed during the relatively brief history of humanity, we know of only about 3,000 that have been used for food. Today we cultivate only about 150 plant species, nine of which—wheat, rice, corn, barley, sorghum/millet, potato, sweet potato/yam, sugar cane, and soybean—contribute three-quarters of our plant-derived food energy.

The plants in this exclusive club are the result of thousands of years of selection by nature as well as by humans attempting to grow food. Those plants whose genetic endowment allowed them to survive and reproduce passed on their genes to their descendants, while their less hardy relatives perished. Genetic variation due to the natural process of mutation and cross-breeding of the best crop plants by farmers dating back to the Stone Age has led to plants often adapted to a wide range of climate extremes as well as to the constant attacks by pests and diseases.

For example, rice may flourish from sea level to elevations as high as 7,000 feet, or from moderately wet conditions to being submerged in water. Potatoes are grown from the Arctic Circle to southern Africa, from below sea level to 14,000 feet. As a result of millennia of selection, crossing, and cultivation, we have literally thousands of *landraces*, genetically diverse representatives of the world's basic crops, each adapted to its local conditions.

All of our crops are derived from plants that, of course, once grew in the wild, long before humans thought to plant and harvest. In many instances, descendants of their wild ancestors still exist, as well as many closely related species. All of these are a treasure trove for the plant breeder.

The crops grown on the vast acreages of the industrial North have been selected primarily for their ability to produce high yields of useful material—such as grain, seed, leaves, and fruit. But the market value of high-yielding crops can be realized only if the plants can grow vigorously, withstanding the steady onslaught of pests and disease. Insects, bacteria, fungi, and viruses undergo their own genetic changes, some of which enable them to attack crops that lack the gene combinations needed to raise defenses against them.

In nature there is a constant interplay of predator and prey, in which genes determine which will have the upper hand. When a particular crop *cultivar* (cultivated variety) is found to be particularly susceptible to predators such as insects, plant viruses, or molds, the plant breeder needs a source of new genes to incorporate into the cultivar, with the hope that they will confer resistance to the predator.

Sometimes favorable changes in the plant's genes may be achieved by deliberately inducing mutations. For example, genes mutated by X rays have resulted in several useful varieties of rice. Typically, however, the genes needed must be found in other plants. Sometimes this means crossing the plants with closely related varieties. If they are not available, the breeder turns to landraces, by far the biggest source of gene variety. Without the appropriate landrace, the next appeal would be either to a closely related wild species or to the wild ancestors of the crop, if they still exist.

All of this constant, painstaking, lengthy, and costly intermingling of genes from the available gene pool is necessary to stay at least one step ahead of our crops' natural enemies. In the process, breeders of virtually all our modern cultivated crops have had recourse to wild relatives. For example, wild tomato species have been the source of resistance to at least 19 disorders of commercially grown tomatoes. Only recently, resistance to a devastating virus attacking rice was finally found in a few wild species and bred into the agricultural variety. And of course, the raw material of genetic engineering are the genes—entities not created in a laboratory but collected from nature's rich array.

The modern practice of mass plantings of *monocultures*—genetically uniform varieties—appears to make perfect economic sense. The plants grow and mature together, so that the grower can harvest them all at once and offer a product that looks and tastes the same. To perpetuate this homogeneity, the farmer is not always forced to seek out new, resistant varieties. By using chemical sprays and powders the organisms set to attack the crop can be kept at bay—at least until the predators develop resistance to the new chemical defenses.

Recent history documents this fragile balance between abundance and disaster. A few specimens of potatoes were gathered from the Andes, their native habitat, and introduced into Spain in 1570, and then to England and Ireland 20 years later. There, isolated from many of their natural pests and diseases, they flourished for many years. In Ireland, the people became almost entirely dependent on the potato for their sustenance. True, three-quarters of the land supported rich fields of wheat, but this was exported to England or handed over to landlords for rent. In 1845 this massive potato monoculture began to succumb to an infestation of a fungus, a devastating blight that met with no resistance from the withering crops.

Over the next five years, several million people died or were forced to emigrate in the face of imminent starvation. Eventually, resistant varieties of potato were bred from the thousands of varieties then still growing in the Andes, as well as in Mexico. Potatoes survived to become a major crop once again—but too late to save a generation of humans whose fate was tied to a single plant.

The English, whose militia had prevented the hungry Irish from seizing the life-saving wheat before it could be exported, would be the unwitting victims, at least in a social sense, of the loss of a quite different monoculture only a few years later. They became a nation of even more dedicated tea drinkers after a fungus destroyed the coffee plantations in Ceylon, India, East Asia, and parts of Africa.

By the late 1960s most of the corn grown in the United States was of a single variety. In 1970 half of the corn crop was lost to another fungal disease, the southern corn leaf blight, which, uncontested, spread northward through the midwestern corn fields at a rate of 90 miles per day. Resistant corn varieties were tracked down and crossed with the surviving plants to create newly resistant cultivars.

In short, only a handful of crops stand between humanity and mass starvation. In a purely pragmatic sense, the many diverse landraces, wild relatives, and remnant populations of wild ancestors are a diverse pool of genes of inestimable worth, not only for the time-honored practice of plant breeding but now for plant genetic engineering as well. It is a great irony that the very practice of massive plantings of a relatively few homogeneous crops has led to the neglect and subsequent disappearance of an alarming number of related varieties—the irreplaceable source of new and useful genes. Plant genetic engineers cannot create new genes. They must search for them among the living relatives of the crops undergoing genetic engineering—and among them are many plants which are now headed towards extinction.

The Treasures of Biodiversity

Only a few of the crops that we have come to associate with the bountiful harvests of the United States are native to these shores. If we were restricted to local plants we would be raising only sunflowers, blueberries, cranberries, pecans, and chestnuts. The corn, beans, tobacco, and squash that colonizers found already being cultivated by the Native Americans had been introduced from Central America and the Caribbean.

The developed nations thrive on crops such as wheat, soybeans, alfalfa, barley, and tomatoes, all of which were introduced from their area of origin—in most cases in what is now the Third World. In terms of native plants, the richer nations of the world are gene-poor, while the poorer nations are gene-rich. The history of conquest and colonization is intimately associated with the gathering of the bounty of the local flora.

In his 1989 book *Gene Hunters: Biotechnology and the Scramble for Seeds*, Calestous Juma details many vivid examples of the search for botanical treasures from distant lands. He points out that as Queen Victoria began her reign, 13,000 species of such exotic flora were being cultivated in the British Isles.

The utilization of plants from the far reaches of the globe has been worth literally billions of dollars to the nations tied to the constant infusion of new genes into their crops. For example, a Turkish landrace of wheat gave American varieties resistance to stripe rust fungus, an improvement that is estimated to be worth $50 million per year. A lone wild tomato variety collected in Peru in 1962 has meant at least $8 million a year to the tomato processing industry because of its high content of soluble solids. In neither case did the country of origin share in the wealth.

The realization that such plants are a valuable natural resource has only recently been recognized by the nations within whose boundaries the plants grow. Setting aside for a moment the other ethical implications of the practice of the sampling of these plants without adequate compensation, one central issue must be addressed—the real danger of plant extinctions.

Regardless of who owns or uses plants for agricultural, or for that matter medicinal, purposes, their value as a source of genetic variability is unquestioned. We have already mentioned that modern monoculture has led to neglect of many plant varieties. For example, the high-yielding varieties of wheat introduced to much of the Third World over the last 30 years, while they solved many food problems, led to the literal disappearance of many of the older bread-wheat varieties. The wild relatives of crops are often regarded as weeds and rigorously eliminated.

Other wild relatives or ancestors of modern crops have the misfortune to grow in areas encroached upon by the inexorable spread of human populations. Political upheaval endangers others. Valuable landraces of upland Afghanistan wheat were destroyed in the recent war. Logging in the Andes threatens the native potatoes, while the overgrazing of domestic goats takes its toll on the local wild tomato species.

Extinctions of plants whose true genetic value we will never know are happening at an alarming rate in the tropics. Warnings about deforestation have become almost a mantra of the 90s, but meanwhile tropical forests are cleared at a rate of over 40 million acres annually. Scientists estimate that as many as 90 percent of the 10 million or so species of plants and animals on this planet live in the tropics. There, some 60,000 plant species may be wiped out in the next three decades unless deforestation is slowed dramatically.

As rising populations, poverty, environmental neglect, and war, all very much related issues, interact in the complex milieu that characterizes the modern human condition, the resource housing life's currency— plants in whose cells and seeds reside the invaluable genes—are rapidly becoming extinct.

Seed Banks

Any discussion of the struggle for the resolution of the global social problems of which living organisms, including people as well as plants, are unwitting victims is beyond the scope of this book. We can, however, summarize the efforts that have been made by scientists to preserve at least some of nature's botanical genetic treasures. Life's currency is being deposited in banks.

One way to conserve wild plants and landraces is to let them flourish where they are, in the wild or on simple farms. Other options are to maintain them in botanical preserves or to disperse them into the horticultural market and hope that enough people will choose to plant them. All of these are now done, with varying levels of success. Ideally, assuming that the other options are being exercised as well, one would like to have a secure repository for all plant varieties where one could be assured of their survival and where one could have access to literally all plant genes.

There actually is an international collection of seed repositories, whose holdings represent only a small fraction of the world's flora. They are variously referred to as *seed banks*, *gene banks*, or *germplasm banks*. The story of the political, personal, and scientific intrigues behind the founding and operation of this assemblage of banks, which are managed by governments, multinational corporations, private seed companies, universities, and private individuals, is variously interpreted, to put it mildly. For example, at least according to Vice President Gore, in his 1992 book, *Earth in the Balance: Ecology in the Human Spirit*:

> The current [seed bank] system is in scandalous condition, with insufficient government attention and money, little coordination between different repositories, grossly inadequate protection and maintenance of national collections, and a missing sense of urgency. . . .

The year after a resolution was passed at the 1972 Stockholm Conference on the Human Environment calling for concerted global action to save crop genetic resources, CGIAR formed the International Bureau for Plant Genetic Resources to collect and conserve seeds through its system of international agricultural research centers. For example, the International Rice Research Institute in the Philippines has specialized in collecting the 100–120 thousand recognized varieties of rice, while the International Maize and Wheat Improvement Center holds the world's central collection of corn varieties.

The United States had already established a National Seed Storage Laboratory in Fort Collins, Colorado in 1958. Here, just as in the other more than one hundred seed banks scattered across the world, the general approach is to dry crop or crop-related seeds gradually and store them at low temperatures. There are now at least four million seed samples in storage worldwide. However, not all seeds survive in the cold, and some varieties die faster than others. Scientists have become increasingly aware that many seeds require specialized treatment. These technical challenges, as well as financial constraints, have led to a situation where Jaap Hardon, head of the Dutch seed bank, says: "There is more genetic erosion in the banks than outside them." Others consider it more appropriate to refer to some seed banks as "seed morgues."

A compendium of criticisms of seed banks from motives to maintenance can be found in *Shattering: Food, Politics, and the Loss of Genetic Diversity*, written in 1990 by Cary Fowler and Pat Mooney. One critical issue here is the role that these seeds and the genes that they represent play, not just as tools for classic plant breeding, but as raw materials for the new biology of plant genetic engineering.

The Third World holds most of the world's plant genetic diversity.

While the seeds in the CGIAR seed banks are freely available to the countries where they were collected, they are also available to bio-technologists, centered mainly in the industrialized countries, who have the technical know-how to use them to create new varieties. The countries from which the genes originate may find themselves as customers for patented seeds containing those very genes, now made into commodities for private profit.

Vandana Shiva, Director of the Research Foundation for Science, Technology, and Natural Resource Policy in Dehra Dun, India, and Associate Editor of the journal *The Ecologist*, has this view of the CGIAR. Writing in *The Ecologist* in April 1990, she labeled their International Bureau for Plant Genetic Resources as "an instrument for the transfer of resources from the South to the North," which has "used Third World resources for the benefit of the industrialized countries." According to Cary Fowler and Pat Mooney: "All along the way, the irascible opponent of Third World countries' influence over their own germplasm donations has been the United States."

There is a growing realization within the Third World of the value of its genetic resources. The adamant resistance to compensating these countries for their seeds is under increasing pressure. A possible answer to this question of equity for Third World countries, as well as to the serious threat to the world's genetic diversity, may come out of a series of meetings that were held in Keystone, Colorado, in 1988, in Madras, India, in 1990, and in Oslo, Norway, in 1991.

The Keystone Center, a nonprofit organization founded in 1975, is funded by foundation, corporate, and individual contributions. It has focused on looking for creative ways to solve difficult environmental, scientific, and natural resource problems. In this case, activists who had long been critics of seed companies, representatives of those companies, government scientists, and heads of seed banks developed over a three-year period a consensus which was published in 1991 as the "Global Initiative for the Security and Sustainable Use of Plant Genetic Resources."

Their recommendations (which were supported by CGIAR) called for a new international structure for preserving plant germplasm. They urged that $300 million eventually be spent each year to maintain samples preserved in seed banks, to pay farmers to grow traditional varieties that could not be stored, and for international coordination, education, training, and research. The money, donated by all countries, would be kept in a trust fund managed by the World Bank.

The Keystone Center recommendations became an integral part of the Biological Diversity Treaty presented to the delegates to the June 1992 Earth Summit—the United Nations Conference on Environment and Development, held in Rio de Janeiro. The United States was the only nation attending the Earth Summit that did not sign the treaty.

The Bush administration's decision not to sign was reported to have been strongly influenced by a memorandum written by David M. McIntosh, the executive director of the now-defunct President's Council on Competitiveness. He warned that the treaty made it unclear how biotechnology companies would be compensated for products developed from the genes of organisms collected in foreign countries. Former President Bush, in his own words, objected to the "concessional terms for sharing benefits derived from biological resources" and said they "would interfere with the basis of our free trade economy—the system of patent and intellectual rights."

Apparently no other country shared his concern to the same extent. It is not yet clear whether the interpretation of the language of the treaty will limit industry's hold on biological patents or simply guarantee that developing countries receive a fair share in profits from the use of genes from species collected there. One clause which raised strenuous objections from President Bush and his staff held that biotechnologies developed from the resources of one country by a company from another should be shared "on a fair and equitable basis."

It is clear that the Clinton–Gore administration has a different attitude about what went on in Rio. According to President Clinton, "Rather than opposing the efforts made there by other countries, we should have helped shape, and then signed" the treaty. On June 4, 1993, the biodiversity treaty was signed by Madeleine Albright, the American representative to the United Nations.

Whatever the fate of the treaty's positions over the next few years, the United States will most certainly be led by an administration with a different philosophy toward conservation. Vice President Gore writes in his *Earth in the Balance* that

The single most serious strategic threat to the global food system is the threat of genetic erosion: the loss of germplasm and the increased vulnerability of food crops to their natural enemies. . . . It is this supply of genes which is now so endangered.

In a later chapter he goes on to call for a "new Green Revolution that will focus on the needs of the Third World's poor" and says that "elements of a just solution . . . can lead to . . . a powerful and effective global effort to link guarantees of justice for the dispossessed to the granting of financial assistance and technology transfers. . . ."

*　　　　　*　　　　　*

There is little doubt that transgenic plants will become an increasingly important part of the agriculture of the future. But will the scenario of "justice for the dispossessed" expressed by Vice President Gore come to pass as more and more genes are identified, isolated, and used to transform plants into novel, useful, and profitable forms?

Certainly, the science of plant genetic transformation appears to have a bright and fascinating future. However, its contribution beyond the laboratory to leading us toward a safe and abundant food supply, not only for the already privileged but for the world's poor as well, depends on far more than science alone. The challenge of fulfilling the promises of a fair sharing of productive food biotechnology is magnified by the complexities of the new world of the 90s and beyond.

Will expanding populations, depletion of natural resources, and political, social, and economic upheaval move the industrialized nations further toward their own short-term self-interest? Or was the Rio Earth Summit in 1992 a sign of hope that protection of the environment and economics can now be understood as essential partners in sustainable development?

Particularly in the United States, an acknowledged leader in plant biotechnology, the choices of our personal, corporate, and national priorities over the next decade will play a major role in determining whether or not the world moves toward increasing harmony and sustainable agricultural productivity for all.

EPILOGUE

Given the accelerating pace of gene research, gene discovery, and the ensuing applications of new genetic knowledge, a book about our Gene Future must inevitably be written even as the early stages of that very future unfold. But the supply of genes at our disposal now, and the uses to which we put them, is minuscule compared to the cornucopia of genes and genetic manipulations that will become available to us as we greet the next century.

Each day witnesses new discoveries. Over the last two years, Dr. Craig Venter and his colleagues at NIH have identified over 8,000 human genes. In July 1992 Dr. Venter left NIH, taking about 30 NIH research scientists with him, and resumed his computer-assisted search for human genes at the Institute for Genomic Research (IGR) in Gaithersburg, Maryland.

Dr. Venter, along with his staff of 70 scientists, expects to determine the sequence of DNA bases of 2,000 to 3,000 human genes per week. IGR will provide rights to any products made possible by this unprecedented outpouring of gene information to Human Genome Sciences, Inc., a company formed by the Health Care Investment Corporation of Edison, New Jersey. According to an interview in the August 1992 *Genetic Engineering News*, Dr. Venter, referring to similar major programs for gene sequencing in Europe and Japan, remarked that "pretending that there is no scientific competition in this field is foolish. There will be a gene race, and it will be fantastic."

In that same issue, Viren Mehta, an analyst of pharmaceutical stocks, referring to the fact that the bold step to undertake such an ambitious sequencing effort will attract venture capital funds, advised that "What's more important is finding the right horses [genes] to invest in."

As we have seen, organisms that have been genetically modified, such as transgenic animals, can be patented. Now, even fragments of genes are under consideration as patentable. In late 1992 the Patent and Trademark Office (PTO) rejected a controversial application by Craig Venter to patent partial sequences of human genes whose functions are not yet known. Venter's first such application, claiming nearly 2,400 segments from human genes, was filed in conjunction with NIH when Venter was still working there. It was also turned down by the PTO. NIH plans to appeal the decision. Spurred by the specter of losing control over its own discoveries, the Medical Research Council in London filed an application with the U.K. Patent Office for more than 1,000 fragments of human DNA sequences. The Council made it clear that it did not think that anyone should have patent rights to gene sequences of unknown function.

When the NIH patenting effort was first announced, it was widely regarded, in the words of a *Nature* editorial, as "an affront to common sense." NIH responded by explaining that it had filed the patent applications as a preventive measure. NIH officials say they feel that, because NIH routinely publishes its gene sequence information and intends to continue doing so (as does Venter), failing to file could jeopardize the possibility of obtaining patents later, when the function of the genes are known.

Reactions are mixed. The Association of Biotechnology Companies has approved the application process with reservations, but the Pharmaceutical Manufacturers Association is opposed to it and the Industrial Biotechnology Association is expected to do the same. In the uproar which followed the NIH proposal, James Watson, who is highly critical of the patenting efforts, resigned his post as director of the National Center for Human Genome Research, where he headed the Human Genome Project.

Meanwhile, Daniel Cohen and his colleagues at the Center for the Study of Human Polymorphism in Paris unveiled their plan, already well under way, to construct a map of the human genome. Cohen, in an interview in the February 8, 1993, *Time*, criticizes the NIH plans as "like trying to patent the stars." He adds that "By patenting something without knowing the use of it, you inhibit industry. That could be a catastrophe." The Center will donate its gene map to the United Nations, to be freely accessible to all scientists.

The patent issue is by no means resolved. The future will reveal whether or not what has long been assumed to be our common human heritage—our human genes—will someday be considered private property.

Reviewing the prospectus which we have painted for the future in the context of current newspaper headlines, we find that genes and the DNA from which they are fashioned are finding a permanent place in our lives. They have entered our vocabulary, research laboratories, medical centers, courts, farms, fields, and factories. They are being put to use as tools both to seek knowledge and to apply that knowledge in technologies that carry with them both the promise of rewards and, sometimes, the dangers of excesses. For example:

Russian forensic scientist Pavel Ivanov, working with Britain's Home Office, has used DNA fingerprinting to identify the bones of Czar Nicholas II and his family. Bones believed to be those of Russia's last imperial family, executed by the Bolsheviks in 1918, were discovered in the late 1970s. Minute amounts of mitochondrial DNA were extracted from the bones and compared to mitochondrial DNA isolated from a blood sample donated by Prince Philip, the present Duke of Edinburgh. His grandmother and Czarina Alexandra's grandmother were sisters. Philip's DNA matched that of the skeletons—consistent with those believed to be the czarina and three of her daughters.

A group of anthropologists have teamed up with geneticists in planning a concerted effort to reconstruct the history of the human past. They hope to begin, in 1994, a worldwide survey of the genes in hundreds of indigenous populations. They story of human origins and our evolution, diversity, and patterns of migration over the centuries lies hidden for now in the telltale patterns of our DNA, passed down from generation to generation.

The anthropologists intend to study, among others, the Bushmen of South Africa, the African Pygmies, the Hill People of New Guinea, and the Yanomami Indians of the Amazon rain forest—small, isolated populations that only rarely, if ever, have mingled their genes with those of their neighbors. Each of these peoples is a "window into the past," according to population geneticist Kenneth Kidd of Yale. Their DNA will give us a glimpse at the genes of humans who lived thousands of years ago. Blood samples will be taken from at least 25 volunteers from each of the 500 or so groups to be studied, and the samples will be maintained as living cells—permanent reservoirs of DNA for analysis. This will be the first step in building what will amount to a guide to the family tree of all humanity.

In December 1992 Germany's state ethics panel approved one of the first gene therapy trials in Europe. Germany has stringent gene therapy regulations, in response to widespread public concerns over the possible eugenic uses of human gene manipulation. A cancer patient's skin cells will be given interleukin-2 genes (found in normal human immune systems). The cells will then be mixed with some of the cancer cells, irradiated, and injected into the patient as a vaccine in hopes of stimulating production of cancer-fighting white blood cells. The European Community does not yet have a clear format for regulation and approval of gene therapy studies.

Without first consulting the Recombinant DNA Advisory Committee (RAC), former NIH Director Bernadine Healy gave a San Diego doctor permission to try gene therapy on a 51-year-old woman dying of brain cancer. The therapy, very much like that carried out in the first NIH brain gene therapy trial, began on January 4, 1993. An uproar followed Healy's decision, described by her as one based on "compassion," in part because all of the usual painstaking reviews of proposed gene therapy had been disregarded.

Another concern was the context in which the approval had been given. Dr. Healy had been urged by Senator Tom Harkin to approve the treatment. She initially deferred, but reversed her position on December 28, 1992. The fact that Senator Harkin chairs the appropriations subcommittee that oversees the NIH budget raised accusations that political pressure had played an unwelcome role in a complex medical and ethical decision.

The RAC met on January 14 and recommended a protocol for internal NIH review and approval for genetic treatments for dying patients if the RAC could not meet quickly enough to evaluate the procedures. Critics warned of a flood of requests from patients and their physicians to attempt unproven gene therapies.

The People's Republic of China has announced that it has been growing transgenic tobacco. The Chinese have genetically modified the plants to resist the tobacco mosaic virus. According to Zhang-Liang Chen, head of the National Laboratory of Protein Engineering and Plant Genetic Engineering at Peking University, China will use these plants to make tobacco for cigarettes that will be sold beginning in 1994. Consumer trials are already under way. The cigarettes were originally given the name "Gene," but smokers objected—so the name has been changed to "China."

The tobacco test fields, along with fields of disease-resistant transgenic tomatoes and potatoes, apparently are the largest in the world. Zhang-Liang, in remarking about the reticence of Western nations to release transgenic organisms into the environment, said, "If I'm poor, like most of our 1.2 billion people, I think about what I need to eat tomorrow."

As the former Soviet Union struggles with the enormous economic burden attendant upon the collapse of the Communist system, more than half a million varieties of 2,500 plant species stored in seed banks in Eastern Europe and Russia are at risk. Most of the long-term storage facilities holding these precious genetic resources have run out of funding.

Many of these seed types must be regrown and the new seeds dried every five years—a labor-intensive process for which there is little governmental support. Meanwhile, scientists from Russia's renowned Vavilov Institute of Plant Diversity in St. Petersburg have revealed how some of their predecessors starved to death during the Second World War rather than eat the seeds in the Institute's repository. Rather than consume these precious genetic resources, at least nine scientists died during the siege of Leningrad in 1941–1942 while watching over their store of several tons of seeds. After that bitter winter, which damaged many seeds, scientists from the Institute planted as many as possible to try to renew their collections while artillery shells were still falling on the surrounding area.

*   *   *

The New Biology is here to stay. Genes, those coiled molecules of DNA whose internal code calls forth an array of proteins that drive the complex chemistry of life, have mutated, combined, and survived through eons of evolution to form millions of living organisms. Most species that have evolved on this planet are now extinct. *Homo sapiens*, the last surviving human species, has, through the agency of our genes, developed a unique intelligence.

Our curiosity and creativity, made possible by an amazing network of nervous tissue fashioned by an as yet unknown number of our genes, has only recently discovered the existence and molecular secrets of those genes. They have become, paradoxically, our servants. We are no longer totally bound by the vagaries of chance in reproduction. We have begun to be inventors of new life forms, limited in their characteristics only by the constraints of our present technology.

But that technology too will evolve. The limitations set on that technology will not be scientific but social, moral, and ethical. We will be called on to balance the needs of a rapidly expanding population with protection of the environment that supports all of life. We will literally be asked to decide what forms of life we should create through the dicing and splicing of genes.

We will have the opportunity to diagnose, treat, and prevent diseases that have brought suffering into the lives of so many. In doing so we will collect the most intimate data about our personal genetic endowments. This information may be used to safeguard our health and that of future generations. This record of our genes also could be misused unless we can ensure that access to it is limited to legitimate purposes.

Our Gene Future will give us the tools to rebuild and refine life as no other series of discoveries ever has. We can only hope that our human intelligence and creativity, which have amassed so much knowledge, will be supported by the wisdom to make the choices that will work for the common good of all life.

SELECTED READINGS

Chapter One

Berg, Paul, and Maxine Singer, *Dealing with Genes: The Language of Heredity* (Mill Valley, CA: University Science Books, 1992).

Bluestone, Mimi, "Hope and Hype in Biotechnology," *Bio/Technology* (September 1992), pp. 946–948.

Brownlee, Shannon, and Joanne Silberner, "The Age of Genes," *U.S. News & World Report* (4 November 1991), pp. 64–76.

Edgington, Stephen M., "PCR: Catching the Next Wave," *Bio/Technology* (February 1992), pp. 137–140.

Green, Rochelle, "Tinkering with the Secrets of Life," *Health* **22** (January 1990), pp. 46–86.

Hodgson, John, "Sequencing and Mapping Efforts in 'Model Organisms,'" *Bio/Technology* (July 1992), pp. 760–761.

Kennedy, Max J., "The Evolution of the Word 'Biotechnology,'" *Tibtech* (July 1990), pp. 218–220.

Lee, Thomas F., *The Human Genome Project: Cracking the Genetic Code of Life* (New York: Plenum, 1991).

Roberts, Leslie, "Two Chromosomes Down, 22 to Go," *Science* **258** (2 October 1992), pp. 28–30.

Watson, James D., Michael Gilman, Jan Witkowski, and Mark Zoller, *Recombinant DNA: Second Edition* (New York: W. H. Freeman and Co., 1992).

Winnacker, Ernst L., "A Room with a View," *Biotechnology Education* **1**:101–108 (1990).

Chapter Two

Allman, William F., "Who We Were," *U.S. News & World Report* (16 September 1991), pp. 53–60.

Cann, Rebecca L., Mark Stoneking, and Allan C. Wilson, "Mitochondrial DNA and Human Evolution," *Nature* **235** (1 January 1987), pp. 31–36.

Day, Stephen, "The First Gene on Earth," *New Scientist* (9 November 1991), pp. 36–40.

Gibbons, Ann, "Systematics Goes Molecular," *Science* **251** (22 February 1992), pp. 872–874.

—— "Mitochondrial Eve: Wounded, But Not Dead Yet," *Science* **258** (14 August 1992), pp. 873–875.

Gould, Stephen Jay, "Eve and Her Tree," *Discover* (July 1992), pp. 32–33.

Harrison, Richard G., "Animal Mitochondrial DNA as a Genetic Marker in Population and Evolutionary Biology," *Tree* (1 January 1989), pp. 6–11.

Hagelberg, Erika, Ian C. Gray, and Alec J. Jeffreys, "Identification of the Skeletal Remains of a Murder Victim by DNA Analysis," *Nature* **352** (1 August 1991), pp. 427–429.

Hoppe, Kathryn, "Brushing the Dust off Ancient DNA," *Science News* (24 October 1992), pp. 280–281.

Hughes, Austin L., "DNA from a 7000-year-old Brain," *Tree* (12 December 1988), pp. 314–316.

Lewin, Roger, "The Biochemical Route to Human Origins," *Mosaic* (Fall 1991), pp. 46–55.

Margulis, Lynn, and Ricardo Guerrero, "Kingdoms in Turmoil," *New Scientist* (23 March 1991), pp. 46–50.

Shreeve, James, "Madam, I'm Adam," *Discover* (June 1991), p. 24.

—— "The Dating Game," *Discover* (September 1992), pp. 76–83.

Templeton, Alan R., "Human Origins and Analysis of Mitochondrial DNA Sequences," *Science* **255** (7 February 1991), p. 737.

Thorne, Alan G., and Milford H. Wolpoff, "The Multiregional Evolution of Humans," *Scientific American* **266** (April 1992), pp. 76–79.

Vigilant, Linda, *et al.*, "African Populations and the Evolution of Human Mitochondrial DNA," *Science* **253** (27 September 1991) pp. 1503–1507.

Wilson, Allan C., and Rebecca L. Cann, "The Recent African Genesis of Humans," *Scientific American* **266** (April 1992), pp. 66–73.

Chapter Three

Aldhous, Peter, "Congress Reviews DNA Testing," *Nature* **351** (27 June 1991), p. 684.

Anderson, Christopher, "FBI Gives in on Genetics," *Nature* **355** (20 February 1992), p. 663.

—— "FBI Attaches Strings to Its DNA Database," *Nature* **357** (25 June 1992), p. 618.

Annas, George J., "Setting Standards for the Use of DNA-Typing Results in the Courtroom—The State of the Art," *The New England Journal of Medicine* **326** (11 June 1992), pp. 1641–1644.

Brookfield, John, "Law and Probabilities," *Nature* **355** (16 January 1992), pp. 207–208.

Brown, Phyllida, " 'Foolproof' DNA Fingerprints within Grasp," *New Scientist* (23 November 1991), p. 14.

Erikson, Deborah, "Do DNA Fingerprints Protect the Innocent?" *Scientific American* **265** (August 1991), p. 18.

Forensic DNA Analysis (Washington, D.C.: U.S. Government Printing Office, 1992).

Franklin-Barbajosa, Cassandra, and Peter Menzel, "The New Science of Identity," *National Geographic* (May 1992), pp. 112–124.

geneWatch, vol. 8 (April 1992).

Goldberg, Stephanie B., "A New Day for DNA?" *ABA Journal* (April 1992), pp. 84–85.

Howlett, Rory, "DNA Forensics and the FBI," *Nature* **341** (21 September 1989), pp. 182–183.

Jeffreys, Alec J., *et al.*, "Hypervariable 'Minisatellite' Regions in Human DNA," *Nature* **314** (7 March 1985), pp. 67–73.

—— "Positive Identification of an Immigration Test-Case Using Human DNA Fingerprints," *Nature* **317** (31 October 1985), pp. 818–819.

Kirby, L.T., *DNA Fingerprinting: An Introduction* (New York: Stockton Press, 1990).

Lander, Eric, "DNA Fingerprinting on Trial," *Nature* **339** (15 June 1989), pp. 501–505.

—— "DNA Fingerprinting: Science, Law, and the Ultimate Identifier," in Daniel J. Kevles and Leroy Hood, eds., *The Code of Codes* (Cambridge, MA: Harvard University Press, 1992).

McGourty, Christine, "New York State Leads on Genetic Fingerprinting," *Nature* **341** (14 September 1989), p. 90.

Moss, Debra Cassens, "Free at Last," *ABA Journal* (October 1989), p. 19.

National Research Council Report, *DNA Technology in Forensic Science* (Washington, D.C.: National Academy Press, 1992).

Risch, Neil J., and B. Devlin, "On the Probability of Matching DNA Fingerprints," *Science* **255** (7 February 1992), pp. 717–720.

Roberts, Leslie, "Fight Erupts over DNA Fingerprinting," *Science* **254** (20 December 1991), pp. 1721–1723.

—— "DNA Fingerprinting: Academy Reports," *Science* **256** (17 April 1992), pp. 300–301.

—— "Science in Court: A Culture Clash," *Science* **253** (7 August 1992), pp. 732–736.

Shapiro, Martin M., "Imprints on DNA Fingerprints," *Nature* **353** (12 September 1991), pp. 121–122.

U.S. Department of Justice, Federal Bureau of Investigation, *Proceedings of the International Symposium on the Forensic Aspects of DNA Analysis* (Washington, D.C.: U.S. Government Printing Office, 1989).

Vogel, Shawna, "The Case of the Unraveling DNA," *Discover* (January 1990), pp. 46–47.

Chapter Four

Caskey, C. Thomas, "DNA-Based Medicine: Prevention and Therapy," in Daniel J. Kevles and Leroy Hood, eds., *The Code of Codes* (Cambridge, MA: Harvard University Press, 1992), pp. 112–135.

Collins, Francis S., "Cystic Fibrosis: Molecular Biology and Therapeutic Implications," *Science* **256** (8 May 1992), pp. 774–779.

Cox, David R., "Medical Genetics," *Journal of the American Medical Association* **266** (15 July 1992), pp. 368–369.

Diamond, Jared, "The Cruel Logic of Our Genes," *Discover* (November 1989), pp. 72–78.

Fox, Jeffery L., and Jennifer Van Brunt, "Towards Understanding Human Genetic Diseases," *Bio/Technology* (October 1990), pp. 903–909.

Galloway, John, "Cancer Is a Genetic Disease," *New Scientist* (18 March 1989), pp. 54–58.

Joyce, Christopher, "Physician, Heal Thy Genes," *New Scientist* (15 September 1990), pp. 53–56.

Kiester, Edwin, Jr., "A Bug in the System," *Discover* (January 1991), pp. 70–76.

Koshland, Daniel E., "The Cystic Fibrosis Gene Story," *Science* **245** (8 September 1989), p. 1029.

Lewis, Ricki, "Genetic Imprecision," *Bioscience* (May 1991), pp. 288–293.

Marx, Jean L., "The Cystic Fibrosis Gene Is Found," *Science* **245** (1 September 1989), pp. 923–925.

Marx, Jean L., *et al.*, "Biotech's Second Generation," *Science* **256** (8 May 1992), pp. 766–815.

Rousseau, Francis, *et al.*, "Direct Diagnosis by DNA Analysis of the Fragile X Syndrome of Mental Retardation," *The New England Journal of Medicine* **325** (12 December 1991), pp. 1673–1681.

Wallace, Douglas C., "Mitochondrial Genetics: A Paradigm for Aging and Degenerative Diseases?" *Science* **256** (1 May 1992), pp. 628–632.

Chapter Five

Angier, Natalie, "Many Americans Say Genetic Information Is Public Property," *New York Times* (September 29, 1992), Section C, p. 3.

Bonnicksen, Andrea, "Genetic Diagnosis of Human Embryos," *Hastings Center Report Special Supplement* **22** (July–August 1992), pp. S5–S10.

Draper, Elaine, "Genetic Screening: Social Issues of Medical Screening in a Genetic Age," *Hastings Center Report Special Supplement* **22** (July–August 1992), pp. S15–S18.

geneWatch, vol. 6 (April 1990), Special Issue on Human Genetics.

Goleman, Daniel, "New Storm Brews over Whether Crime Has Roots in Genes," *New York Times* (September 15, 1992), Section C, p. 5.

Harper, Peter S., and Angus Clarke, "Should We Test Children for 'Adult' Genetic Diseases?" *The Lancet* **335** (19 May 1990), pp. 1205–1206.

Johnson, Mary, "Defective Fetuses and Us," *The Disability Rag* (March/April 1990), p. 34.

Lacayo, Richard, "Nowhere to Hide," *Time* (11 November 1991), pp. 35–40.

Marteau, Theresa, "The Need for Caution on Genetic Screening," *New Scientist* (15 December 1990), p. 6.

Morris, Michael J., *et al.*, "Problems in Genetic Prediction for Huntington's Disease," *The Lancet* (9 September 1989), pp. 601–605.

Nelkin, Dorothy, "The Social Power of Genetic Information," in Daniel Kevles and Leroy Hood, eds., *The Code of Codes* (Cambridge, MA: Harvard University Press, 1991), pp. 177–190.

Nelkin, Dorothy, and Laurence Tancredi, *Dangerous Diagnostics: The Social Power of Biological Information* (New York: Basic Books, 1989).

Nolan, Kathlen, "First Fruits: Genetic Screening," *Hastings Center Report Special Supplement* **22** (July–August 1992), pp. S2–S4.

Orentlicher, David, "Genetic Screening by Employers," *Journal of the American Medical Association* **263** (16 February 1990), pp. 1005–1008.

—— "Use of Genetic Testing by Employers," *Journal of the American Medical Association* **266** (2 October 1991), pp. 1827–1830.

Platt, Lawrence D., and Dru E. Carlson, "Prenatal Diagnosis—When and How?" *The New England Journal of Medicine* **327** (27 August 1992), pp. 636–638.

Vines, Gail, "For Good or Evil: Genetic Tests for the Masses," *New Scientist* (11 August 1990), pp. 54–55.

Chapter Six

Anderson, W. French, "Genetics and Human Malleability," *Hastings Center Report* **20** (January–February 1990), pp. 21–24.

—— "Human Gene Therapy," *Science* **256** (8 May 1992), pp. 808–813.

—— "Uses and Abuses of Human Gene Transfer," *Human Gene Therapy* **3**:1–2 (1992).

Angier, Natalie, "Doctors Have Success Treating a Blood Disease by Altering Genes," *New York Times* (July 28, 1991), Section 1, p. 1.

Beardsley, Tim, "Profile: Gene Doctor," *Scientific American* **263** (August 1990), pp. 33–35.

Begley, Sharon, and Daniel Shapiro, "The Death of the 'Bubble Boy,'" *Newsweek* (5 March 1984), p. 71.

Berger, Edward M., and Bernard M. Gert, "Genetic Disorders and the Ethical Status of Germ-Line Gene Therapy," *The Journal of Medicine and Philosophy* **16**: 667–683 (1990).

Collins, Francis S., "Cystic Fibrosis: Molecular Biology and Therapeutic Implications," *Science* **256** (8 May 1992), pp. 774–779.

Fletcher, John C., "Evolution of Ethical Debates about Human Gene Therapy," *Human Gene Therapy* **1**:55–68 (1990).

Friedmann, Theodore, "The Evolving Concept of Gene Therapy," *Human Gene Therapy* **1**:175–181 (1990).

Friedman, Theodore, and Richard Roblin, "Gene Therapy for Human Genetic Disease?" *Science* **175** (3 March 1972), pp. 949–955.

Golde, David W., "The Stem Cell," *Scientific American* **265** (December 1991), pp. 86–93.

Jaroff, Leon, "Giant Step for Gene Therapy," *Time* (24 September 1990), pp. 74–76.

Juengst, Eric T., "The NIH 'Points to Consider' and the Limits of Gene Therapy," *Human Gene Therapy* **1**:425–433 (1990).

Kolberg, Rebecca, "U.S. Alters Cold Virus to Treat Cystic Fibrosis," *New Scientist* (12 December 1992), p. 5.

Larrick, James W., and Kathy L. Burck, *Gene Therapy : Application of Molecular Biology* (New York: Elsevier, 1991).

Miller, Dusty, "Human Gene Therapy Comes of Age," *Nature* **357** (11 June 1992), pp. 455–460.

Montgomery, Geoffrey, "The Ultimate Medicine," *Discover* (March 1990), pp. 60–68.

Nicholas, Eve K., ed., *Human Gene Therapy* (Cambridge, MA: Harvard University Press, 1988).

Office of Technology Assessment, U.S. Congress, "Human Gene Therapy," OTA-BP-BA-32: (Washington, D.C.: Government Printing Office, 1984).

"Points to Consider in Human Somatic Cell Therapy and Gene Therapy," *Human Gene Therapy* **2**:251–256 (1991).

Tauer, Carol A., "Does Human Gene Therapy Raise New Ethical Questions?" *Human Gene Therapy* **1**:411–418 (1990).

Thompson, Larry, "At Age 2, Gene Therapy Enters a Growth Phase," *Science* (30 October 1992), pp. 744–746.

Verma, Inder M., "Gene Therapy," *Scientific American* **263** (November 1990), pp. 68–84.

Chapter Seven

Barinaga, Marcia, "Knockout Mice Offer First Animal Model for CF," *Science* **257** (21 August 1992), pp. 1046–1047.

Begley, Sharon, "Barnyard Bioengineers," *Newsweek* (9 September 1991), p. 55.

Cherfas, Jeremy, "Molecular Biology Lies Down with the Lamb," *Science* **249** (14 July 1990), pp. 124–126.

Ditullio, Paul, *et al.*, "Production of Cystic Fibrosis Transmembrane Conductance Regulator in the Milk of Transgenic Mice," *Bio/Technology* (January 1992), pp. 74–77.

Gordon, Jon W., and Frank H. Ruddle, "Integration and Stable Germ-Line Transmission of Genes Injected into Mouse Pronuclei," *Science* **214** (11 December 1981), pp. 1244–1246.

Gordon, Jon W., *et al.*, "Genetic Transformation of Mouse Embryos by Microinjection of Purified DNA," *Proceedings of the National Academy of Science* (December 1980), pp. 7380–7384.

Greave, David R., *et al.*, "A Transgenic Mouse Model of Sickle Cell Disorder," *Nature* **243** (11 January 1990), pp. 183–185.

Hanson, Betsy, and Dorothy Nelkin, "Public Responses to Genetic Engineering," *Society* (November–December 1989), pp. 76–80.

Jaenisch, Rudolf, "Transgenic Animals," *Science* **240** (10 June 1988), pp. 1468–1474.

Larrick, James W., and Kathy L. Burck, *Gene Therapy: Application of Molecular Biology* (New York: Elsevier, 1991).

Lewis, Sherry M., and Jefferson H. Carraway, "Large Animal Models of Human Disease," *Lab Animal* (January 1992), pp. 22–29.

Loew, Franklin M., "Animal Agriculture," in *The Genetic Revolution: Scientific Prospects and Public Perceptions* (Baltimore: Johns Hopkins University Press, 1991).

MacDonald, June Fessenden, ed., *Animal Biotechnology: Opportunities & Challenges* (Ithaca, NY: National Agricultural Biotechnology Council, 1992).

"Man's Mirror," *The Economist* (16–22 November 1991), pp. 21–24.

Mark, Willie H., "Manipulation of the Mouse Genome Using Embryonic Stem Cells," *Lab Animal* (July–August 1992), pp. 41–53.

Merlino, Glenn T., "Transgenic Animals in Biomedical Research," *The FASEB Journal* (November 1991), pp. 2996–3001.

Mestel, Rosie, "The Mice without Qualities," *Discover* (March 1993), pp. 18–19.

Moffat, Anne Simon, "Transgenic Animals May Be Down on the Pharm," *Science* **254** (4 October 1991), pp. 35–36.

Mroczek, Nancy S., "Point of View: Recognizing Animal Suffering and Pain," *Lab Animal* (October 1992), pp. 27–31.

National Academy of Sciences, The, "Genetic Engineering Is Environmentally Safe," in (William Dudley, ed.), *Genetic Engineering: Opposing Viewpoints* (San Diego: Greenhaven Press, 1990), pp. 32–38.

Padres, Herbert, Anne West, and Harold Alan Pincus, "Physicians and the Animal-Rights Movement," *The New England Journal of Medicine* (6 June 1991), pp. 1640–1643.

Pursel, Vernon G., *et al.*, "Genetic Engineering of Livestock," *Science* **244** (16 June 1989), pp. 1281–1288.

Rollin, B.E., "The 'Frankenstein Thing': The Moral Impact of Genetic Engineering of Agricultural Animals on Society and Future Science," in (Steven M. Gendel *et al.*, eds.), *Agricultural Bioethics; Implications of Agricultural Biotechnology* (Ames, IA: Iowa State University Press, 1990), pp. 292–308.

Saffer, Jeffery D., "Transgenic Mice in Biomedical Research," *Lab Animal* (March, 1992), pp. 30–38.

Sharples, Frances E., "Research in Genetic Engineering Must Proceed," in (William Dudley, ed.), *Genetic Engineering: Opposing Viewpoints* (San Diego: Greenhaven Press, 1990), pp. 39–45.

Spalding, B.J., "Transgenic Pharming Advances," *Bio/Technology* (May 1992), pp. 498–499.

Thomas, Kirk R., and Mario R. Capecchi, "Site-Directed Mutagenesis by Gene Targeting in Mouse Embryo–Derived Stem Cells," *Cell* (6 November 1987), pp. 503–512.

U.S. Congress, Office of Technology Assessment, "Emerging Animal Technologies," in *A New Technological Era for American Agriculture*, OTA-F-474 (Washington, D.C.: U.S. Government Printing Office, 1992).

Watson, James D., Michael Gilman, Jan A. Witkowski, and Mark J. Zoller, "The Introduction of Foreign Genes into Mice," in *Recombinant DNA* (New York: W. H. Freeman and Co., 1992).

Watts, Susan, "A Matter of Life and Patents," *New Scientist* (12 January 1991), pp. 56–61.

Chapter Eight

Brunke, Karen J., and Ronald L. Meeusen, "Insect Control with Genetically Engineered Crops," *Tibtech* (June 1991), pp. 197–200.

Buck, Kenneth, "Brave New Botany," *New Scientist* (3 June 1989), pp. 50–56.

Day, Stephen, "A Shot in the Arm for Plants," *New Scientist* (9 January 1993), pp. 36–40.

Dodds, John H., and Lorin W. Roberts, *Experiments in Plant Tissue Culture: Second Edition* (Cambridge, U.K.: Cambridge University Press, 1985).

Dhir, Sarwan K., *et al.*, "Regeneration of Transgenic Soybean (*Glycine max*) Plants from Electroporated Protoplasts," *Plant Physiology* (1992), pp. 81–87.

Evans, David A., and William R. Sharp, "Applications of Somaclonal Variation," *Bio/Technology* (June 1986), pp. 528–532.

Fromm, Michael E., Loverine P. Taylor, and Virginia Walbot, "Stable Transformation of Maize after Gene Transfer by Electroporation," *Nature* **139** (27 February 1986), pp. 791–793.

Gasser, Charles S., and Robert T. Fraley, "Genetically Engineered Plants for Crop Improvement," *Science* **245** (16 June 1989), pp. 1293–1299.

Hamilton, A.J., G.W. Lycett, and D. Grierson, "Antisense Gene that Inhibits Synthesis of the Hormone Ethylene in Transgenic Plants," *Nature* **346** (19 July 1990), pp. 284–287.

Klein, Theodore M., *et al.*, "Transformation of Microbes, Plants, and Animals by Particle Bombardment," *Bio/Technology* (March 1992), pp. 286–291.

Kung, Shain-dow, and Ray Wu, *Transgenic Plants: Engineering and Utilization* (San Diego: Academic Press, 1993).

——*Transgenic Plants: Present Status and Social and Economic Impacts* (San Diego: Academic Press, 1993).

Lindsey, Keith, and M. G. K. Jones, *Plant Biotechnology in Agriculture* (Englewood Cliffs, NJ: Prentice Hall, 1990).

Moffat, Anne Simon, "Engineering Useful Plants," *Mosaic* **21** (Spring 1990), pp. 34–43. "Improving Plant Disease Resistance," *Science* **254** (24 July 1992), pp. 482–483.

Neumann, Eberhard, and Petra Bierth, "Gene Transfer by Electroporation," *American Biotechnology Lab* (March–April 1986), pp. 11–14.

Oeller, Paul W., *et al.*, "Reversible Inhibition of Tomato Fruit Senescence by Antisense RNA," *Science* (18 October 1991), pp. 437–439.

Oxtoby, Elli, and Monica A. Hughes, "Engineering Herbicide Tolerance into Crops," *Tibtech* (March 1990), pp. 61–65.

Pool, Robert, "In Search of the Plastic Potato," *Science* **245** (15 September 1989), pp. 1187–1189.

Porier, Yves, *et al.*, "Polyhydroxybutrate, a Biodegradable Thermoplastic, Produced in Transgenic Plants," *Science* **256** (24 April 1992), pp. 520–524.

Raineri, D. M., *et al.*, "*Agrobacterium*-mediated Transformation of Rice (*Oryza sativa* L.), *Bio/Technology* (January 1990), pp. 33–38.

Rhodes, Carol A., *et al.*, "Genetically Transformed Maize Plants from Protoplasts," *Science* **240** (8 April 1988), pp. 204–207.

Savitz, Eric J., "A Taste of the Future," *Barron's* (1 June 1992), pp. 8–24.

U.S. Congress, Office of Technology Assessment, *A New Era of Technology for American Agriculture*, OTA-F-474 (Washington, D.C.: U.S. Government Printing Office, 1992).

Chapter Nine

Anderson, Christopher, "Researchers Ask for Help to Save Key Biopesticide," *Nature* (20 February 1992), p. 661.

Arntzen, Charles J., "Regulation of Transgenic Plants," *Science* **257** (4 September 1992), p. 1327.

"Biotechnology and the American Agricultural Industry," *Journal of the American Medical Association* **265** (20 March 1991), pp. 1429–1436.

Burros, Marian, "Gene-spliced Foods: Is It Safe Soup Yet?" *New York Times* (June 17, 1992), Section C, p. 3.

Canine, Craig, "Brave New World: Keeping the Promise of Biotechnology," *Harrowsmith Country Life* (September–October 1990), pp. 36–43.

Charles, Dan, "White House Changes Rules for Genetic Engineering," *New Scientist* (25 May 1991), p. 14.

Davis, Bernard, and Lissa Roche, "Genetic Engineering: Sorcerer's Apprentice or Handmaiden to Humanity?" *USA Today* (November 1989), pp. 68–70.

Doebley, John, "Molecular Evidence for Gene Flow among *Zea* Species," *Bioscience* (June 1990), pp. 443–448.

Doyle, Jack, "Biotechnology's Bitter Harvest of Herbicides," *geneWatch* **2**: 1–19 (1985).

Dudley, William, ed., "Does Genetic Engineering Improve Agriculture?" *Genetic Engineering: Opposing Viewpoints* (San Diego: Greenhaven Press, 1990), pp. 114–163.

Ellstrand, Norman C., and Carol A. Hoffman, "Hybridization as an Avenue of Escape for Engineered Genes," *Bioscience* **40** (June 1990), pp. 438–442.

Flavell, Richard B., *et al.*, "Selectable Marker Genes: Safe for Plants?" *Bio/Technology* (February 1992), pp. 141–144.

Fox, Jeffery L., "USDA Snarls at Transgenic Catfish," *Bio/Technology* (May 1992), p. 492.

—— "USDA Eases Field-Testing Rules," *Bio/Technology* (December 1992), p. 1524.

Fraley, Robert, "Sustaining the Food Supply," *Bio/Technology* (January 1992), pp. 40–42.

Gussow, Joan Dye, *Chicken Little, Tomato Sauce, and Agriculture: Who Will Produce Tomorrow's Food?* (New York: Bootstrap Press, 1991).

Hoffman, Carol A., "Ecological Risks of Genetic Engineering of Crop Plants," *Bioscience* **40** (June 1990), pp. 434–437.

Hoyle, Russ, "Eating Biotechnology," *Bio/Technology* (June 1992), p. 629.

—— "FDA's Slippery Food Policy," *Bio/Technology* (September 1992), pp. 958–959.

—— "Rifkin Resurgent," *Bio/Technology* (November 1992), pp. 1406–1407.

Huttner, Susanne L., *et al.*, "Revising Oversight of Genetically Modified Plants," *Bio/Technology* (September 1992), pp. 967–971.

Keehn, Joel, "Mean Green," *Buzzworm: the Environmental Journal* (January–February 1992), pp. 33–37.

Keeler, Kathleen H., "Can Genetically Engineered Crops Become Weeds?" *Bio/Technology* (November 1989), pp. 1134–1139.

Kessler, David A., *et al.*, "The Safety of Foods Developed by Biotechnology," *Science* **256** (26 June 1992), pp. 1747–1748.

Keystone National Biotechnology Forum, *An Analysis of the Federal Framework for Regulating Planned Introductions of Engineered Organisms* (Keystone, CO: Keystone Center, 1989).

Lambert, Bart, and Marnix Peferoen, "Insecticidal Promise of *Bacillus thuringiensis*," *Bioscience* (February 1992), pp. 112–122.

McCormick, Douglas, "Frankenfood . . . or Frank Discussion?" *Bio/Technology* (August 1992), p. 829.

McDonald, June Fessenden, ed., *Agricultural Biotechnology, Food Safety, and Nutritional Quality for the Consumer*, National Agricultural Biotechnology Council Report 2 (Binghamton, N.Y.: Union Press, 1990).

Miller, Julie Ann, "Biosciences and Ecological Integrity," *Bioscience* **41** (April 1991), pp. 206–210.

Pimental, David, "Down on the Farm: Genetic Engineering Meets an Ecologist," *geneWatch* **4**(May–June 1987), pp. 8–11.

"Proposed Guidelines for Research Involving the Planned Introduction into the Environment of Organisms with Deliberately Modified Hereditary Traits; Notice," *Federal Register* (1 February 1991), pp. 4135–4143.

Rennie, John, "Putting Down Roots," *Scientific American* **262** (May 1990), pp. 81–84.

"Statement of Policy: Foods Derived From New Plant Varieties: Notice," *Federal Register* (29 May 1992), pp. 22984–23005.

Tiedje, James M., *et al.*, "The Planned Introduction of Genetically Engineered Organisms: Ecological Considerations and Recommendations," *Ecology* (April 1989), pp. 298–315.

U.S. Congress, Office of Technology Assessment, *New Developments in Biotechnology: Field-Testing Engineered Organisms: Genetic and Ecological Issues*, OTA-BA-350 (Washington, D.C.: U.S. Government Printing Office, 1988).

Ward, K.A., and C. D. Nancarrow, "The Genetic Engineering of Production Traits in Domestic Animals," *Experientia* (1991), pp. 913–922.

Wrubel, R. P., S. Krimsky, and R. E. Wetzler, "Field Testing Transgenic Plants," *Bioscience* **42** (April 1992), pp. 280–289.

Chapter Ten

Adams, Robert P., and Janice E. Adams, eds., *Conservation of Plant Genes: DNA Banking and in Vitro Biotechnology* (San Diego: Academic Press, 1992).

Board of Agricultural Staff, eds., *Managing Global Genetic Resources: The U. S. National Plant Germplasm System* (Washington, D.C.: National Academy Press, 1991).

Brown, Lester R., "Feeding Six Billion," *World Watch* (September–October 1989), pp. 32–40.

Bunders, Joske F. G., ed., *Biotechnology for Small-Scale Farmers in Developing Countries* (Amsterdam: VU University Press, 1990).

Cohen, Joel J., *et al.*, "Ex Situ Conservation of Genetic Resources: Global Development and Environmental Concerns," *Science* **253** (23 August 1991), pp. 866– 871.

Dalberg, Kenneth A., "Sustainable Agriculture—Fad or Harbinger?" *BioScience* (May 1991), pp. 337–340.

Gibbons, Ann, "Saving Seeds for Future Generations," *Science* **254** (8 November 1991), p. 804.

Goodman, M. M., "Genetic and Germ Plasm Stocks Worth Conserving," *Journal of Heredity* (1990), pp. 11–15.

Goldburg, Rebecca, *et al.*, "Biotechnology's Bitter Harvest: Herbicide-Tolerant Crops and the Threat to Sustainable Agriculture," Report of the Biotechnology Working Group, Tides Foundation, March 1990.

Gore, Albert, *Earth in the Balance: Ecology and the Human Spirit* (Boston: Houghton Mifflin Company, 1992).

Harlander, Susan K., "Biotechnology—A Means for Improving our Food Supply," *Food Technology* (April 1991), pp. 84–144.

Hathaway, C. Michael, "Yes: A Threat to Property Rights," *ABA Journal* (September 1992), pp. 42–43.

Hindmarsh, Richard, "The Flawed 'Sustainable' Promise of Genetic Engineering," *The Ecologist* (September–October 1991), pp. 196–205.

Hogdson, John, "Biotechnology: Feeding the World," *Bio/Technology* (January 1992), p. 4750.

Hoyle, Russ, "Winning The Tomato War," *Bio/Technology* (December 1992), pp. 1520–1521.

Juma, Calestous, *The Gene Hunters: Biotechnology and the Scramble for Seeds* (Princeton, NJ: Princeton University Press, 1989).

Kats, Robert W., *et al.*, "The Future of Hunger," *Occasional Papers: World Hunger Program, Brown University* (14 February 1988), pp. 23–45.

Kirschenmann, Frederick, "Green vs. Gene," *Agricultural Engineering* (January 1992), pp. 22–24.

Kloppenburg, Jack, Jr., "No Hunting!" *Cultural Survival Quarterly* (Summer 1991), pp. 14–18.

Lubchenco, Jane, *et al.*, "The Sustainable Biosphere Initiative: An Ecological Research Agenda," *Biology* (1991), p. 371.

Messer, Ellen, and Peter Heywood, "Trying Technology: Neither Sure Nor Soon," *Food Policy* (August 1990), pp. 336–345.

"Miracle or Menace for the Third World?" *Geographical Magazine* (January 1991), pp. 40–43.

Nestle, Marion, "Food Biotechnology: Truth in Advertising," *Bio/Technology* (September 1992), p. 1056.

Persley, Gabrielle J., *Beyond Mendel's Garden: Biotechnology in the Service of World Agriculture* (Tucson, AZ: CAB International, 1990).

Pimental, David, *et al.*, "Benefits and Risks of Genetic Engineering in Agriculture," *Bioscience* (October 1989), pp. 606–614.

Reganold, John P., Robert I. Papendick, and James F. Parr, "Substainable Agriculture," *Scientific American* **261** (June 1990), pp. 112–120.

Sattaur, Omar, "The Shrinking Gene Pool," *New Scientist* (29 July 1989), pp. 37–43.

Shiva, Vandana, "Biodiversity, Biotechnology and Profit: The Need for a People's Plan to Protect Biological Diversity," *The Ecologist* (April 1990), pp. 44–47.

Tudge, Colin, *Food Crops for the Future* (New York: Basil Blackwell, Inc., 1988).

Weintraub, Pamela, "The Coming of the High Tech Harvest," *Audubon* (July–August 1992), pp. 92–103.

Wilson, Edward O., "Threats to Biodiversity," *Scientific American* **261** (September 1989), pp. 108–116.

Platais, Kerri Wright, and Michael P. Collinson, "Biotechnology and the Developing World," *Finance & Development* (March 1992), pp. 34–36.

Young, Robert M., ed., *Science as Culture* (London: Free Association Books, 1991).

Epilogue

Coghlan, Andy, "China's New Cultural Revolution," *New Scientist* (2 January 1993), p. 4.

Eisenberg, Rebecca S., "Genes, Patents, and Product Development," *Science* **257** (14 August 1992), pp. 903–907.

Fox, Jeffery L., "PTO Nixes NIH's DNA Patents," *Bio/Technology* (November 1992), p. 1410.

"Gene Patents," *Nature* **359** (1 October 1992), p. 348.

Roberts, Leslie, "How to Sample the World's Genetic Diversity," *Science* **257** (28 August 1992), pp. 1204–1205.

Thompson, Larry, "Harkin Seeks Compassionate Use of Unproven Treatments," *Science* **258** (11 December 1992), p. 1728.

—— "Healy Approves an Unproven Treatment," *Science* **259** (8 January 1993), p. 172.

—— "Should Dying Patients Receive Untested Genetic Methods?" *Science* **259** (22 January 1993), p. 452.

INDEX